# A study of so influencing fertilit the bull

Herbert Lester Gilman

**Alpha Editions**

This edition published in 2024

ISBN : 9789364734028

Design and Setting By
**Alpha Editions**
www.alphaedis.com
Email - info@alphaedis.com

As per information held with us this book is in Public Domain.
This book is a reproduction of an important historical work. Alpha Editions uses the best technology to reproduce historical work in the same manner it was first published to preserve its original nature. Any marks or number seen are left intentionally to preserve its true form.

# A STUDY OF SOME FACTORS INFLUENCING FERTILITY AND STERILITY IN THE BULL

## HERBERT L. GILMAN

*Veterinary Experiment Station, Cornell University*

Normal reproduction is the fundamental foundation upon which the entire cattle industry rests. For this reason, any factor capable of interfering with it is a detriment to the industry, and a matter of prime importance to the breeder and the veterinarian. With the relative increase in number and value of cattle, and the fact that the profession is depending more and more on this industry for a livelihood, these problems are assuming greater importance. The part played by the bull has been emphasized entirely too little, with the result that, as in human medicine, many fail to appreciate the effects of sterility or lowered fertility in the male. The part played by the sire in the spread of genital infections, though discussed frequently, has received little systematic investigation.

The bull mast be regarded as at least half the herd, not only from the standpoint of the characters he imprints upon his progeny, but because of his relation to the reproductive efficiency in the herd. It seems quite probable that he does disseminate during copulation, infection associated with the genital organs, with the result that the bull is a very important factor in a study of the subject. Too frequently, his ability to copulate in an apparently normal manner, is taken as a standard of fertility. Gross changes in his genitalia, or the absence of spermatozoa from the semen are given due consideration, while other more obscure abnormalities are not looked for nor regarded in their proper light. Neither fertility nor sterility are always absolute, but the terms should be used relatively inasmuch as we may have all degrees of infertility or impotency. All too frequently we forget the many delicate and intricate mechanisms involved in the reproductive process, with the result that many phases of the problem are neglected or disregarded. The genital organs work as a unit, each part of which must function in perfect accord with the others to the end that full fertility may result. The physiological factors involved in the formation of the semen are too little understood, or at best, our knowledge regarding them is more or less hazy.

The purposes of the present work have been: (1) to summarize the work so far done on the subject, (2) to review briefly the known facts throwing light on the anatomy and physiology of the male genital organs, (3) to carry our systematic studies upon the pathology and bacteriology of the genital tract of the bull, and (4) to ascertain if possible whether the bull is a disseminator of those infections which interfere with reproduction in the female.

The work has been carried on for the most part from the point of view of a laboratory man cooperating with clinicians. No attempt is made in this paper to give detailed clinical data, methods for physical examinations, etc. There are included many statements and some data, given in a preliminary article on the subject. While the subject is broad in its scope, in fact too broad for great detail, it is hoped that a start has been made toward future and more detailed investigations.

# History

References to, and investigations relating to, the part played by the bull in the process of reproduction in the herd, and in the spread of genital infections, have been limited largely to those phenomena caused by *Bact. abortum*. Bang (1) originally called attention to the possibility of the male transmitting the organism discovered by him, but he reached no definite conclusion on the subject. James Law (2) writing on contagious abortion in cows, early suspected this possibility when stating under "casual infections," that—"In a case which came under the observation of the writer recently, a family cow, kept in a barn where no abortion had previously occurred, was taken for service to a bull in a herd where abortion was prevailing, and though she was only present at the latter place for a few minutes, she aborted in the sixth month." Jansen, as quoted by Sand, reports the case of a cow from an aborting herd having been taken into a herd that had been previously quite free from the disease. Soon after her arrival she aborted, and later cow after cow of the original herd aborted. The owner kept the matter a secret, and sent his cow to a neighbor's bull for service, with the result that for two years abortion prevailed among cows served by this bull. McFadyean and Stockman (3) later, in experimental work, attempted but failed to infect cows by using a soiled bull for service. Hadley and Lothe (4) state: "A large number of stockmen hold that the bull is an important factor in the transmission of contagious abortion in herds. A smaller number believe that the bull merely acts as a passive carrier of the abortion disease and is not actively concerned in the transmission." In a subsequent bulletin, Hadley (5) remarks: "The abortion organisms may enter the body ... during sexual intercourse." In an experiment carried on by the same author and co-workers, abortion-free virgin heifers were mated to abortion-infected bulls, infection being evidenced by positive reactions to the complement fixation and agglutination tests. His results indicate, he believes, "that the bull is not so important a factor in transmitting abortion as many believe." The conclusions are: "Bulls may become infected with abortion bacilli. Bulls that reacted to the blood tests were incapable of disseminating the abortion disease to the abortion-free heifers with which they were mated. Bulls appear to possess a sexual or individual immunity to abortion infection that renders them less susceptible than cows and induces a milder form of the disease. The resistance appears to be due to certain anatomic and physiologic differences in their sexual organs which make them less favorable places for the growth of the abortion germs than those of the opposite sex."

Buck, Creech, and Ladson (6) applied the agglutination test to 325 mature bulls, of which 288 were negative and 37 positive. *Bacillus abortus* was isolated from five animals, of which three showed marked lesions, two in the seminal vesicles, and one in the left testicle. They conclude: "*B. abortus* may involve organs of the generative apparatus of bulls, producing chronic inflammatory changes. Of the generative organs, the seminal vesicles appear to furnish the most favorable site for the lodgement and propagation of abortion infection."

Schroeder and Cotton (7) cite the case of a bull which reacted to the abortion test and, on post mortem, *Bact. abortum* was isolated from an abscess of one epididymis. They state: "Our attempts to produce a similar case of infection artificially failed, and, in agreement with the difficulties many investigators have had to obtain incriminating evidence against bulls, we have thus far failed to infect bulls in any way that justifies the assumption that they are important factors in the dissemination of abortion disease." Further, they conclude: "Regarding the dissemination of abortion disease by bulls, we may say, however, that it would be foolhardy in the dim light of our present knowledge to take liberties with reacting bulls, or bulls from infected herds, or promiscuously used bulls."

Cotton (8) failed to demonstrate the presence of abortion bacilli in the genital organs of the bull used to serve aborting cows, or in the testicles of two bull calves, one of which had been fed and the other injected with the cultures of the abortion bacillus. He concludes that the bull does not harbor the organisms in the testicles. Carpenter (9) injected both *streptococci* and *Bact. abortum* into the scrotal sacs of young calves, and intravenously in others. In no case was he able to recover the organisms from any part of the genital canal, except for a *streptococcus* in one instance. Rettger and White (10) were unable to obtain evidence of the presence of *Bact. abortum* in three bulls slaughtered after repeated reactions to the complement fixation and agglutination tests. The three bulls had been under observation for three years, with no conclusive evidence to indicate that they were a source of danger to the herds in which they were a part. They believe that the bull transmits the infection as a passive carrier.

Attempts at artificial inoculation by natural channels have failed, with the possible exception of McFadyean, Sheather, and Minett (11) who were able to infect the bull by the prepuce in two cases and by the mouth in one case. The results, however, are by no means conclusive. They conclude, nevertheless, that cattle of any age of either sex may be infected by natural channels with the bacillus of epizootic abortion.

Schroeder (12) carried out investigations to ascertain the frequency with which bulls react to abortion tests, and the frequency with which lesions chargeable to abortion bacilli occur in the reproductive organs of reacting bulls. Studies were also pursued which he states conclusively prove that bulls with infected reproductive organs may expel abortion bacilli with the seminal fluid. In the first two mentioned investigations 325 bulls from a Washington abattoir were tested, and slaughtered upon reaction. "Approximately ten per cent of the bulls reacted, and approximately ten per cent of the reacting bulls showed lesions of the reproductive organs from which abortion bacilli were isolated." The value of these studies, he emphasizes, lies not in "that they give us a measure of the proportion of bulls that react positively to abortion tests or the proportion of reacting bulls that are carriers of abortion bacilli," but in "the fact that they show that abortion bacillus disease of the bull's reproductive organs is not a wholly unique affection which practically may be ignored, but an important condition that must be taken into account in our efforts to combat infectious abortion, since it has been proved to be associated with contamination of the seminal fluid." In discussing the method by which infected bulls transmit the organisms to cattle, he believes that leakage of semen from the penis, or vaginas of cattle after service, contaminates the food which subsequently gains entrance to their digestive tracts. As the result of a series of experiments, he states: "... the results fail to justify in the least degree the assumption that cows are infected with abortion bacilli via their vaginas or uteruses at the time of copulation, or that the bull, through copulation, is an agent in the spread of abortion disease."

The work so far alluded to, has been limited to infection with, and the transmission of, *Bact. abortum* and the lesions associated with such infection. The last mentioned author, however, states: "A search for other specific causes of abortions among cattle has not been neglected, and bureau investigators could relate at great length stories similar to those which investigators have told about microorganisms isolated from the products of abortions and the uteruses of cows that have aborted. Bacilli of various kinds, different types of micrococci, and spirilla or vibrio have been found repeatedly; but when their pathogenicity has been tested in accordance with widely recognized and accepted and required bacteriological standards, not one shred of evidence has been obtained to prove them true etiological factors of bovine abortions. What role such microorganisms may have as causes of the sequellae of infectious abortions, and of other, possibly, independent, abnormal processes in the reproductive organs, is far from clear and merits careful study." Hadley (5) mentions the fact that: "Unquestionably the male often becomes infected with the germs that produce the various secondary diseases in the female, which are properly classed under the more inclusive term 'abortion disease.'" Also, speaking of the rarity with which the bull acquires abortion infection, he alludes to the fact that he may act as a "mechanical carrier of various disease germs from an infected to a healthy cow." Carpenter (9) working on the female genital tract, comes to the conclusion that, in all probability, the genital organs are normally free from bacteria. Barney (13) quoting Huet finds that bacteria may be present in the seminal vesicles of healthy animals (horses, cattle, pigs, and laboratory animals). This, he states, corresponds with the well recognized findings in other parts of the genito-urinary tract, not only in animals, but in man. He (Huet) has further found that in animals dying of acute septicemia, the specific organism (anthrax, pneumococcus) is to be found in the vesicular secretions. Furthermore it was definitely shown that an infection could be transmitted to the female during the act of copulation.

Williams, W. L. (14), calls attention to the lack of veterinary literature relating to the pathology and bacteriology of the male genital tract, except as related to infection with *Bact. abortum*. Infection with other types of bacteria is emphasized, the clinical recognition of such, with the accompanying pathological changes, and of the numerous phenomena involved in the process of reproduction in the male. The semen and its essential germinal elements are taken up with reference to the entire lack of study devoted to them, and some of the abnormal changes are described. In a later contribution (15), he takes up the part played by the bull in the dissemination of genital infections and states: "Clinical studies now indicate with great clearness that the bull is an active spreader of that group of genital infections which cause sterility, abortion, and related phenomena."

Williams, W. W. (16), studied the semen with reference to sterility, emphasizing the importance of its examination in the diagnosis, giving methods for collecting samples, staining of sperms, and some of the abnormalities encountered. The work is fundamental, and should be of great practical importance to all interested in the problem. In a later paper (17), he brings out a more extended discussion of the question. He concludes that the clinical examination is of vital importance, and that the efficiency of the semen depends not only upon its physical properties but upon the number of spermatozoa that are motile, the degree of motility, degree of obligospermia, and the percentage of imperfect spermatozoa, either deformed or immature. Of forty bulls examined, he finds that twenty, or fifty per cent, showed lessened fertility, and others, aside from this, showed minor changes in the genital organs or semen. The same author subsequently takes up the subject of reproduction from the viewpoint of both sexes, but emphasizing infection in the male, and the frequency with which lowered vitality of the germinal cells occurs. Hopper (18) states: "A diseased bull may manifest non-fertility or decreased potency in different ways—by repeated service to apparently normal females without conception, by a high abortion rate in females that have been apparently normal, by characteristic infections following the use of any particular sire, or by abnormalities in the breeding tract noted by rectal or physical palpation."

The observations of Williams (19) in a pure bred dairy herd bring out quite clearly the relation of the bull to the dissemination of genital infections. The bulls in this particular herd were abnormal in many respects, as demonstrated by pathological changes in their genital organs, bacterial invasion of the parts, abnormalities of the semen and spermatozoa, and the probable transmission of infection to the females. Several of the sires from this herd furnished much of the material for the early basic work of this investigation. Since then the tracts of other sires have been worked upon with quite similar or identical results.

To summarize the work already done, most investigators have considered the bull as merely a mechanical carrier of *Bact. abortum* infection, though all are more or less suspicious of his ability to become an active spreader. Schroeder, however, states that the organisms are eliminated with the semen, but infection of the female occurs secondarily through the digestive tract by contamination of the food with the semen. Other investigators bring out fundamental points demonstrating the importance of other organisms than the Bang bacillus and call attention to the need of a more thorough study of the anatomy, physiology, and pathology of the male genital tract.

Any study of the genital organs must of necessity rest fundamentally upon a thorough knowledge of the anatomy and physiology of those parts. Too few of us have stopped to consider these questions thoroughly, with the result that our ideas on the problem are more or less vague. It is much easier to understand why abnormal spermatozoa occur so frequently, or changes take place in the semen with death or weakening of the germinal elements, if we realize or stop to consider the highly differential process of spermatogenesis, and the various structures which contribute to the formation of the semen. We must come to realize that each part of the genital tract is essential to the normal functioning of the whole, and that the genital tract and reproduction are in turn dependent upon the proper functioning of the entire body.

Walker (20) emphasizes the importance of a thorough knowledge of physiology in stating, "Although the subject of sterility has attracted the attention of the medical profession; and although much has been written on its causes and treatment, it cannot be claimed that the practical results obtained up to the present time are satisfactory, or that when consulted for sterility, the medical man of today can hold out to his patient much more hope of successful treatment than the medical man of fifty years ago. Our failure in this respect is in the main due to an ignorance of the physiology of reproduction."

To bring out some of these points, the anatomy of the tract will be reviewed briefly, together with the physiology of reproduction, and the various factors which should be considered in a study of the problem.

# Anatomy and Physiology

In origin and early development the ovary and testis are identical. The gonad and mesonephros or primitive kidney are developed from the urogenital fold. The gonad first forms as a medio-ventral thickening of the fold, which gradually expands until it becomes attached by a mere stalk. At first, the gland is made up merely of a superficial epithelial layer, and an inner epithelial mass, or epithelial nucleus. In the process of development, large primordial germ cells migrate from the entoderm of the future intestinal canal, and pass through the stalk to the gonad. In the case of the male gonad, seminiferous tubules are very difficult to make out in embryos smaller than 24 millimeters. Then they suddenly differentiate out as solid cords of germ cells, while the connective tissue grows in around them. These connective tissue sheaths unite at the center of the organ to form the anlage of the mediastinum testis. The testicular tubules unite and converge toward the hilus, there to meet the anlage of the rete. At the mesonephric end of the testis, the rete first appears as a collection of cells, differentiating out from the inner epithelial mass of the gonad. These cells gradually grow out to meet the collecting portions of the mesonephric tubules on the one hand, and the seminiferous tubules on the other. The rete is represented as cords of cells at first, which in forty millimeter embryos hollow out to form tubules.

The mesonephros, or primitive kidney, early starts to degenerate cranio-caudally,—the tubules becoming separated into a cranial and caudal group. The collecting and secretory parts of the cranial group separate, the collecting tubules growing out to meet the rete with which they unite to form the efferent ductules of the epididymis. The caudal group of tubules is vestigial and becomes the paradidymis. The mesonephric duct becomes the vas deferens, connecting as it does with the tubules of the epididymis, and emptying into the urethra at Müller's tubercle or, as it later becomes, the colliculus seminalis.

The seminal vesicles arise as hollow saccules from the dorsal wall of the mesonephric duct just as it empties into the urethra. The prostate develops as an outgrowth of the dorsal urethra just posterior to Müller's tubercle. The bulbo-urethral glands appear as solid, paired, epithelial outgrowths from the entoderm of the urogenital sinus.

Müller's duct, at first a solid tube growing from the anterior part of the mesonephros, and ending at Müller's tubercle, becomes a hollow tube, and in the female forms the entire genital tract except for the gonad and the lower part of the vagina. In the male, the anterior part remains as the vestigial appendix testis, and the posterior part, as the vagina masculina. Ellenberger states, however, that this embryonic structure is very seldom seen in the mature bull.

The *Male Reproductive Organs* include the penis and testes, together with the excretory passages which connect the testes with the urethral canal. These excretory ducts include the epididymis, vas deferens, and seminal vesicles. Posterior to their termination in the urethra, there are connected the ducts of the prostate gland and the bulbo-urethral or Cowper's glands.

TESTES: The testicles of the bull are relatively large. Varying with the size and age of the animal they measure from fourteen to seventeen centimeters in length, including the epididymis, and from six to eight centimeters in diameter. Each testicle is enclosed within a serous sac, the tunica vaginalis, whose visceral layer is very intimately fused with the underlying covering of the organ, the tunica albuginea. The tunica albuginea is quite thin and consists of connective tissue which is rich in elastic fibres. Muscular tissue is not present as it is in the case of many mammals. Inside the tunica, and closely attached to, though separated from, the parenchyma by a thin layer of connective tissue, is a layer of very loose connective tissue, which because of its rich supply of blood vessels is termed the tunica vasculosa. The parenchyma is of a yellowish gray color, and of a rather soft consistency. It is made up of the seminiferous tubules, rete, and the connective tissue stroma, the mediastinum testis. On section, the mediastinum appears as the center or axis of the entire organ. It is star-shaped, and radiates connective tissue septa out into the parenchyma to support and separate the tubules. Ellenberger states that the testis of the bull and all ruminants lacks a closed system of interlobular septa, because of the feeble development of the connective tissue.

The principal blood vessels and rete tubules are found in this structure, the function of the latter being to connect the seminiferous tubules and the efferent tubules of the epididymis. The epithelium of the rete is quite irregular,—consisting in places of a single layer; in others, of two layers. At some points there are formed groups of several cells lying over one another, with swollen homogeneous basal cells, which sometimes form projections into the lumen.

The interstitial tissue, besides conveying the blood vessels to the organ, contains many "interstitial cells." These cells are relatively sparse in the adult bull, and are comparatively delicate, slightly granular, often shuttle-shaped, with a rather small nucleus. Embryologically they are derived from a syncitium arising from the mesothelium of the genital ridge, differentiating out by growth of the cytoplasm. They contain large quantities of fat, and elaborate the internal secretion of the testis. This secretion governs the development of the secondary sexual characters, and has a profound influence on the general body metabolism, and development of the skeleton. The interstitial cells appear early in embryonic life even before there is any differentiation of sex, and their greater relative development in the fetus is indicative of a future male development. In very young embryos, the growth is very rapid, followed, however, by a period of atrophy, during which the seminiferous tubules undergo marked development. Pende (21) states: "There seems to be an inverse relation between the growth of the tubular and interstitial tissues, as one is hypoplastic when the other is in full activity." From birth to the onset of sexual maturity, which may be called a period of rest for the testicle, the cells are few in number. With the accentuation of the secondary sexual characters, and the beginning of sexual life, these cells again increase in number and activity.

The parenchyma of the testis consists for the most part of the seminiferous tubules, which, on account of the courses they take in the different regions, are divided into groups. The peripheral tubules are the much-contorted tubuli contorti. These anastomose to form the much shorter tubuli recti. These in turn anastomose frequently, uniting to form the rete testis. The rete proceeds through the mediastinum to form the efferent ductules which break through the tunica albuginea to form the greater part of the head of the epididymis. The tubuli contorti are the longer and more numerous of the tubules, for it is here that practically all the spermatozoa are produced. The straight tubules are relatively so short that they may be regarded more in the light of the beginning of the system of excretory ducts.

The seminiferous tubules consist of a thin peripheral membrana propria upon which rests the seminal epithelium, which is made up of the essential semen forming cells, and the cells of Sertoli. The spermatogenic cells may be divided into three groups, from within outward: the peripheral single layer of small cuboidal spermatogonia; one or two rows of large spermatocytes; and three to five rows of spheroidal spermatids. The cells of Sertoli are more or less of the syncitial type,—large in size and irregular in outline. They occur at various intervals between the layers of spermatogenic cells, with their bases resting upon the membrana propria. Centrally they send out protoplasmic processes for variable distances,—some even reaching the border of the innermost cell layers.

SPERMATOGENESIS: In this process, the primary germinal cells, the spermatogonia, divide to form the primary spermatocytes. Maturation consists of two cell divisions of the primary spermatocytes, and these in turn form four spermatids. During the process, the number of chromosomes is reduced to half the number characteristic of the species.

The spermatids then become converted into mature spermatozoa. This mode of transformation may be seen in Plate I. In the process, the nucleus of the spermatid forms a large part of the head: the centrosome divides, part passing to the extremities of the neck. One centrosome becomes the anterior, and remains attached to the head, while the other divides to form the posterior centrosome. The latter is divided into the anterior part, and the posterior nodule or annular ring. Besides this, the posterior centrosome becomes elongated to form the axial filament, and the cytoplasm forms the sheaths of the neck and tail. The spiral filament of the connecting piece is derived from the cytoplasmic mitochondria. At this time, a large part of the cellular cytoplasm is cast off. Meanwhile, the spermatozoa sink their heads into the long protoplasmic processes of the Sertoli or "nurse" cells which furnish nutritive material for their complete development. Finally the adult cells are cast off into the lumen. The structure of the spermatozoa, and a discussion of the semen will be taken up later.

EPIDIDYMIS: The epididymis is divided into three parts: the head, body and tail. The head is made up principally of the lobules formed by the much-coiled efferent ductules proceeding from the rete. The ductules, about twelve in number in the bull, unite to form the body, which remains coiled and runs along the postero-medial part of the testicle to which it is more or less closely attached. To quote Ellenberger (23): "The transition from the rete into the ductules is gradual, as the characteristic epithelium of the latter (ductules) begins in cavities without walls, and at first, gradually form a wall which is well marked out as a thin ring of interstitial tissue.... The epithelium of the ductules is in sharp contrast to the rete in that it has a single-layered ciliated columnar epithelium, in which here and there one finds round basal cells. The dark and light columnar cells alternate; the cilia are often cemented together, and form cone-shaped, homogeneous appearing protuberances. The secretory activity is quite clearly observable. In the light cells one finds secretory globules, accumulating in rows, sometimes above, other times below, arches of cells. The secretory droplets pass from the cells into the lumen, and often lie in irregular layers on the epithelium; also the basal cells appear swollen and shoved out between the cylindrical cells." At the lower extremity of the testicle, the tail is formed, which is globular in shape, and more or less loosely attached to the testicle. Here the ductules anastomose freely, gradually become less coiled, and end in the single excretory tube, the vas deferens. The epithelium at the tail part is more or less of the pseudo-stratified columnar, ciliated variety. Outside this is a membrana propria, a circular muscular layer, and a connective tissue coat. The secretory activity is very marked here and one finds much secretion in the lumen, Courrier (24), working on the bat, suggests that the glandular activity is conditioned by the secretion from the interstitial (endocrine) gland. The action of the secretion is to dilute the large mass of spermatozoa present, nourish them to some extent and also stimulate them to active motility. Stigler (25) states that the properties of the sperms are modified in the epididymis; the motility, the ability to resist heat, and other properties are augmented, at least in the case of the guinea pig, rat, and mouse. Some authors state that the sperms first become motile when in contact with the prostatic secretion, but I have repeatedly examined the contents of the tail of the epididymis of the bull, rabbit, and guinea pig, finding full motility in each case, though the duration is not nearly as long as when the sperms are ejaculated in the semen.

The VAS DEFERENS is quite narrow (2 mm.) and runs from the tail of the epididymis to the verumontanum, or colliculus seminalis, where it empties into the urethra in common with the duct of the vesicle. At first it is lined by epithelium similar to that of the vas epididymis, but this changes over into a peculiar low stratified type. Ellenberger describes it as follows: "The epithelium shows a very pronounced basal coat. The overlying cell zone shows more (at the most, three) rows lying over each other of elongated nuclei, while an outline of cell form is not ordinarily noticeable, so that it may be spoken of as a syncitium, and at the same time as a many layered epithelium." The mucosa forms low, broad folds into the lumen. The tunica propria is a thin connective tissue layer. The submucosa consists of thin connective tissue. Three muscular coats are present: an inner thin longitudinal layer, middle circular layer, and an outer longitudinal layer. All are more or less intimately blended, and are surrounded by the adventitia, made up of connective tissue, elastic fibres and scattered longitudinal muscle cells of the internal cremaster muscle. Near the dorsal surface of the bladder, the ducts come in close apposition, and for ten to twelve centimeters dilate to form the ampullae. Here the mucous membrane becomes much plicated, forming long folds which anastomose freely. The function of the vas is to convey the spermatozoa and secretions from the epididymis to the urethra. Disselhorst (26) believes the ampulla acts as a seminal reservoir and states that he has found spermatozoa stored up in the little pockets in the walls of this structure in animals during the rutting time. He suggests, further, that there is a relation between the state of development of the ampulla and the time occupied by copulation. When the organ is small or absent, as in dogs, cats, and boars, the coition is a slow process, but when the ampulla is large and well developed, as in horses and sheep, the coitus requires a relatively short time. Inasmuch as coitus is so rapid in the bull, and the ampulla is so well developed, it seems as though this function is very probable.

The SEMINAL VESICLES are very compact glandular structures lying on either side of the median line, on the dorsal side of the bladder, and ventral to the rectum. In the mature bull they measure ten to twelve centimeters in length, four centimeters in width, and about two and one-half to three centimeters in thickness. The glands are distinctly lobulated, quite tortuous, and are often asymmetrical in size and shape. They converge posteriorly, to empty into the urethra at the colliculus seminalis with the ampulla, in a slightly oval slit in the mucosa. Microscopically, the gland is of the anastomosing tubular type, with very poorly developed excretory ducts to the glandular cavities. Posteriorly one finds centrally a few sinus-like narrow excretory passages, which open into the somewhat larger collecting and excretory duct. The epithelium is of the simple columnar type and produces a relatively large amount of secretion. The gland cavities are surrounded by a membrana propria, over which is a relatively thick layer of smooth muscle. Outside this is a connective tissue covering which sends trabeculae or septa in between the lobules. The secretion of the seminal vesicles is a tenacious albuminous fluid with a slightly yellowish tinge, all or part of which appears in the ejaculate in the form of swollen sago-like grains which are soon dissolved following ejaculation and the liquefaction of the semen. The proteid compounds belong to the group of histones. The secretion is liquid when warm and coagulates when cold. Some say that the filling of the vesicles serves to excite sexual feeling, but this is doubtful in view of the fact that in some animals the sexual desire exists before the vesicles are filled. Likewise, Steinach found that rats, whose seminal vesicles had been removed, still retained their desire for copulation. The function of the secretion is to furnish much of the volume to the semen, and in some way it has a distinct bearing on fertility, inasmuch as extirpation of the organs in rats leads to lowered fertility. The vesicles of the bull are in no sense a store-house for spermatozoa, as is usually understood. Repeated examinations in a large number of bulls have led to the finding of spermatozoa there only in very rare instances. That they serve as a resorption place for sperms that are not ejaculated is also very unlikely. Normally, one sees on smear of the vesicles, occasional cells, leucocytes, lecithin granules, sago bodies, and rarely a few degenerated spermatozoa.

The COLLICULUS SEMINALIS is a rounded or cone-shaped eminence in the posterior urethra, upon which the ducts of the seminal vesicles and vasa deferentia open. The ducts open separately at the bottom of two narrow slits, one on each side of the mound, there being no distinct ejaculatory duct as in man. The function of the colliculus or verumontanum is not definitely known. It is generally believed that the structure is made up of blood spaces which become engorged during erection, causing a blockage of the posterior urethra, which prevents regurgitation of the semen. Rytina (27), however, demonstrated that the structure is not composed of any unusual number of blood vessels or spaces, and that removal of the organ was not followed by regurgitation of the semen into the bladder during ejaculation. He believes, and quite logically, that its function is to afford a prominence upon which the ducts may empty. The mixture of the thick gelatinous semen with the thin prostatic secretion must occur at the moment of ejaculation and must be perfectly homogeneous, otherwise large numbers of the organisms remain in the thick gelatinous portion of the fluid. The eminence serves this purpose in that the prostatic ducts which converge toward it, may eject their secretion toward the eminence, producing an admixture more evenly and quickly.

PROSTATE: The bull possesses what Ellenberger calls a diffuse prostate. That is, there is no distinct glandular body as in man. It is composed principally of a glandular sheath around the urethral wall. Just posterior to the neck of the bladder, and in front of the urethral muscle, there is formed a slight dorsal transverse elevation, extending downward on the sides. This is what might be termed the body. The greater part of the gland is "disseminate" in form, being a sheath of glandular tissue embedded in the urethral wall. Dorsally it is about ten to twelve millimeters thick, and ventrally about two millimeters. The gland is a branched tubular structure, the interlobular tissue of which contains much unstriped muscle. The lobules are lined by a columnar type of epithelium. The ducts, about thirty to forty in number, open into the urethra in two rows posterior to the colliculus. The secretion is a thin, slightly turbid fluid, of a faintly alkaline reaction. Its function is to dilute the semen, stimulate the motility of the spermatozoa and nourish them.

Fish (28) believes that the activating property of the secretion is due to enzymes, because boiling deprives the fluid of its power to accelerate the motility of the spermatozoa. Serrlach and Pares, quoted by Marshall (29), working on dogs, have adduced evidence indicating that the prostate is an internal secretory gland which controls the testicular functions, and regulates the process of ejaculation. It is stated that if the prostate is removed, spermatozoa are no longer produced in the testis, and that the secretory activity of the accessory genital glands ceases. These changes, however, can be prevented by the administration of extracts of the prostate. The fact that the prostate elaborates a secretion having a definite relation to the testis, has been verified by other authors.

COWPER'S GLANDS (Bulbo-Urethral): These glands are paired, oval structures about one and one-half by two and one-half centimeters in size, situated on either side of the dorsal pelvic part of the urethra close to the ischial arch. They are deeply embedded, with the bulbus urethae, in the bulbo-cavernosus muscle, thereby being inaccessible to palpation. In general, they are of a well developed anastomosing tubular type. The connective tissue stroma is relatively thin, and thickens only in between the larger lobules, where one finds the larger collecting ducts. Each gland opens by a single duct. The epithelium is of the simple cuboidal type. Little is known of the function of its secretion, though Kingsbury (30) regards it in the light of a mating gland; that is, it lubricates the genital passages during coitus, as does its homologous structure in the female, Bartholin's gland.

Ellenberger describes the urethra, slit ventrally, as presenting the following picture: "The colliculus seminalis distinctly appears as a process or offshoot of the crista urethralis of the Trigonium Vesicae. At the summit, and at the bottom of the two slits, open laterally the ducts of the vesicles, and medially the ductus deferens.... From the caudal slope of the colliculus go two distinct mucous membrane folds which run through the entire pelvic urethra, near together, somewhat diverging, and then coming together, so that they form an elongated, narrow oval. At their caudal union, the excretory ducts of the bulbo-urethral glands open side by side. At the point of departure of the folds from the colliculus, arises a niche-shaped opening, between both folds, and likewise lateral to each fold. In these niches open the ducts of the prostate. The openings of the disseminate prostate lie in rows as in the horse, but form not less than six rows. There is one row medial to each fold, and two lateral. Mullet mentions only the medial rows. These rows extend to the opening of the ducts of Cowper's glands. The stratum glandulare (disseminate prostate) is very easy to recognize with the naked eye."

SEMEN: The semen is the mixed product of the testicles, their excretory passages, and the accessory sexual glands, a fact which complicates its study considerably. The freshly ejaculated fluid is cloudy, tenacious, more or less coagulable, and is rich in albumen. It is weakly alkaline in reaction, and contains eighty to ninety per cent of water. Of the solid constituents, there is forty per cent of ash, of which three-fourths is calcium phosphate. Besides the spermatozoa, the semen frequently contains epithelial cells, leucocytes, concentric amyloid concretions, and lecithin bodies. When cold, the characteristic phosphoric acid salts are precipitated. The fluid content is the product of the tubules of the testicles, their excretory ducts, and the accessory sexual glands. The characteristic odor of the semen is supplied by a slimy nucleoalbumen "spermin" which forms the spermatic crystals, and is furnished by the prostatic secretion.

During ejaculation the spermatozoa and secretions added by the testicle and epididymis are probably carried to the ampulla by peristaltic muscular action, in the earlier stages of the orgasm. At the height of the orgasm, the ampulla is emptied into the posterior urethra in common with the secretion of the contracting vesicles, here to be admixed with the thin prostatic secretion. The entire mixture is then propelled, and ejaculation produced by strong muscular contractions of the entire urethra. As was stated before, the semen is the product of the testicles, the excretory ducts, and the accessory sexual glands. The testes furnish the essential germinal elements, the sperms, and some of the fluid content. Then is added the product of the epididymis, vas, and ampulla, which stimulates the spermatozoa to active motility, nourishes the organisms, and adds somewhat to the fluid bulk of the secretion.

Stigler (25) states: "During its sojourn in the epididymis, the properties of the spermatozoa are modified; the motility, ability to resist heat, and other properties are augmented, at least in the case of the guinea pig, rat, and mouse." To the secretion is then added the product of the vesicles, which contributes markedly to its fluid content, nourishes the sperms, and supplies the ferment which induces clotting of the semen when ejaculated. This is very important because the clot formed in the vagina protects the delicate spermatozoa from the hostile acid vaginal secretion. The prostate likewise adds bulk and nourishing substances, besides stimulating the spermatozoa to fuller and more lasting motility. The addition of the spermin is perhaps unimportant. The function of the secretion of the Cowper's glands, which is added at this time, is problematical. It does, however, have a diluting action on the semen. Perhaps its secretion is poured out prior to ejaculation so as to lubricate the canal and prepare the way for the semen.

Fish (28) has demonstrated by means of darkfield illumination, the presence of numerous minute particles or ultraparticles in this fluid. Their character and significance are matters of conjecture, but it would seem as though they were not identical with the "chylomicrons" or fat particles found in the blood by Gage. Perhaps further researches will reveal some intimate connection between the number present in a field, and the relative potency of the animal.

Each portion of the tract furnishes some essential element to the mixed product which is so remarkably adapted both as a vehicle for the ejaculation of the spermatozoa, and as a fluid in which their motility is initiated and maintained. Any derangement of one part is fraught with danger to the existence of viable spermatozoa, and the continuation of full fertility on the part of the animal. The physiology of each contributing gland must be borne in mind at all times. Walker (31) investigated the fertility of the semen of the dog, taken from various parts of the tract. His results were: (1) semen from the testicle and head of the epididymis showed no motility, (2) semen from the tail of the epididymis showed some motility in the more fluid contents of the preparations, (3) semen from the vas deferens appeared about the same, (4) a mixture of epididymis semen and prostatic secretion showed active motility, and (5) likewise in a mixture of epididymis semen, though only in those places where the fluids had become well mixed. My observations, however, differ in one respect with regard to the bull, as I have found full motility of the spermatozoa from the epididymis, but it is not so lasting as when augmented by the addition of the prostatic fluid. Boettcher (32) concludes: "that the secretion of the accessory male genital organs possesses a protective colloid, which (1) hinders the spermatozoid action of the vaginal secretion, at least until the sperms have time to reach the interior of the uterus which is an alkaline reaction, (2) that it makes the ejaculate more voluminous, so that by cohabitation, a very good part of the vagina becomes moistened, and the spermatozoa become distributed over the greater part of the vaginal mucosa. This distribution is rendered necessary because some of the fluid flows from the vagina following coitus. In this manner the opportunity is given for a part of the ejaculate containing the spermatozoa to be brought easier to the external os, and (3) it happens that because of its content of sodium chloride the life of the spermatozoa is stimulated and prolonged." A fuller discussion of the essential physiology of the various parts of the tract on the semen content, and fertility, will be taken up later. The changes in biochemical content, reaction, and the result of the addition of bacterial products will also be more fully discussed.

SPERMATOZOA: The history of the discovery of spermatozoa is very interesting, and for that reason a brief outline will be given. The "semen threads" were first observed in the year 1667, by Ham, a student of Leeuwenhoek at Leyden. The discovery was announced, confirmed by findings in the dog and rabbit, and discussed by the latter author under the title: Observationes Anthonii L. de natis e semine genitali animaculis (Upon the formation of young from procreative material). The sperms were taken to be animals on account of their motility, and their significance remained questionable if not unknown. Spallanzani, quoted by Marshall (29), was the first to show that the filtered fluid was impotent, and that spermatozoa in the semen were essential to fertilization. Kolliker, in 1841, discovered that the sperms arise from the cells of the testis, and Barry in 1843, observed the conjugation of sperm and ovum in the rabbit. This led to a clear understanding of the function of these important germinal elements.

The spermatozoa are the male procreative cells, and are characterized by the possession of a head containing the chromosomes necessary for fertilization, and a tail capable of propelling the organism on its way to meet the ovum. The length of the entire sperm, including the head, is seventy-five to eighty microns. The head is nine and five-tenths microns long, and five and five-tenths microns wide. It may be divided into two principal parts, the head and tail. The head, for the larger part, is made up of the nucleus, and may be differentiated by staining reactions into a darker staining posterior part, an anterior lighter part, and often a still lighter area between the two. On the anterior part is a sharpened edge, the acrosome, which serves to perforate the ovum. The whole is surrounded by a very definite limiting membrane which often becomes obscured under abnormal conditions. The tail may be divided into three parts: connecting piece, principal part, and terminal filament. The connecting piece, the essential motile apparatus, is the thickest and strongest part, and joins the tail proper to the head. It consists of the central axial filament, a spiral filament around this, and an outer mitochondrial covering. Anteriorly it is limited by the anterior portion of the posterior chromosome, and posteriorly by the annular chromosome. The anterior chromosome is directly connected with the head, there appearing a light break here at the neck where separation frequently occurs. This neck serves as a joint for the motion. The axial filament, therefore, does not reach the head, but extends back from the anterior part of the posterior chromosome. The principal part consists merely of the axial filament, and a thin outer covering, while the end piece is quite thin and is made up solely of the uncovered axial filament. The finer structures are seen only when special staining reactions are used, and then only when the sperms are obtained directly from the testicle. The function of the sperms is of course primarily that of fertilization. Numerous observers have, however, thought that they might have some other definite, though unknown, use.

An editorial in the Journal of the American Medical Association (33) raises several important questions regarding this obscure phenomenon. The fact that an enormous number of spermatozoa are produced, and only one, or at most, a few perform the function of fertilization, raises the question as to what becomes of the remainder. It is stated: "Zoologists have found that in some of the invertebrates the spermatozoa invade the entire body of the female, and in some species they reach the ovum by penetrating the cuticle from outside and migrating to their goal. Studies on rodents show that the sperms invade the epithelium of the generative mucosa and underlying connective tissue. These tissues seemed to be stimulated to growth, suggesting that this may influence the uterine mucosa in its preparation for receiving and embedding the egg, and in forming the decidua." It has been shown that the sperms contain a specific protein capable of producing antibodies in the blood plasma, by citing the fact that rabbits develop a distinct Aberhalden reaction for testicular proteins shortly after cohabitation. One very important observation showed that by immunizing female rabbits with sperms they were rendered sterile for some time, although after a few months they again became capable of impregnation. The question raised is: "... if the spermatozoa invade the female tissues and cause the formation of specific antibodies which are capable of preventing fertilization, may not such a process participate in the problem of sterility?" This very problem seems to be a factor in explaining why some couples who are not fertile to each other subsequently are both fertile when they cohabit with other individuals.

*Motility*: After clinical observation of the motility of the spermatozoa of the bull, I find that it differs little or none from the types as observed by Reynolds (34) in his work on human spermatozoa. His observations are so accurate and well described that they will be given in his words. "All normal motions appear to be consecutive phases. Initial motion, i. e., motion as seen in fresh semen under favorable conditions, consists of a lashing of the after part of the tail from side to side which is so rapid as to constitute vibration. It produces rapid forward motion in a practically straight line, the head, middle piece, and forward portion of the tail maintaining their position in the line of motion with practically no swaying from side to side. The action of the flagellum is so rapid that it is quite impossible to follow its individual movements. Spermatozoa swimming in this manner head against the current and usually cross the field of observation in about five seconds in the absence of currents or obstacles. This type of motion will be described throughout the paper as 'progressive vibratile' motion. Progressive vibratile motion is normally succeeded after a variable length of time by what I regard as the second phase of normal motion.

"The second normal motion differs from the first not only in its character but in markedly reduced speed. The tail movement alters to a long stroke from side to side and almost the whole length of the tail partakes in the stroke. This is, moreover, accompanied by swaying of the head and middle piece through an arc which is always considerable and may even equal ninety degrees. The general outline of the spermatozoa, from being practically straight with almost non-detectable sharp, quick, small arc vibration of the after-tail, has become an S in outline, with large, slow, plainly perceptible undulations traveling gradually backward throughout the length of the spermatozoon. Speed has been lost and direction seems to be more specifically determined by the surroundings. Individuals at this stage show a pronounced choice of direction and go up to objects in the medium, from which they later make off as though the movement were determined by tactile reaction to some extent. This type of motion has, therefore, been named 'undulatory tactile' in contradistinction to 'progressive vibratile.'

"The third type of normal motion succeeds the second and consists in a tendency on the part of the spermatozoon to push itself against or into any small masses of cells, or sometimes other materials, which it may find in the neighborhood, bunting itself into any small cove that can be found, and maintaining a slight burrowing motion by a lashing tail movement of the vibratile type not unlike the movements of the caudal fin of a fish. The movement of the flagellum in this third type is unlike the second type in that it is vibratile rather than lashing, but is slower than the vibratile motion of the first type and less limited to the afterpart of the tail. These spermatozoa are apparently not caught in the debris or unable to move off. From time to time, they back out of such a cove and seek another mooring place.

"This 'stationary hunting' motion is less universal than the other two. Many individual spermatozoa fail to attain it. It seems probable that only the most vigorous individuals ever reach this stair". It has not been encountered in unmixed semen or in any artificial mediums. It has been observed only when the spermatozoon is in the secretions of the female genital tract. It is most frequent when the spermatozoon is in contact with a nest of epithelial cells....

"The three types of normal motion are not only distinctive but are always consecutive, i. e., the second follows the first after a period which is apparently determined both by initial vitality and by the favorable or unfavorable character of the medium, while the third has been observed to occur only in individuals which have already developed the second. They apparently constitute a normal cycle.

"This cycle is open to the theoretical explanation that these types of motion are directly adapted to the function of the spermatozoon; thus, the progressive vibratile motion which is characteristic of the earliest period of its existence appears especially suitable for its prolonged journey through the cervix and uterus to the fundus and tube. This is supported by the fact that during this motion it always heads directly against any existing current, and that during this stage of its journey it must under natural conditions be continuously exposed to the faint outward ciliary currents of the mucous membrane of these passages.

"The undulatory tactile motion which succeeds the progressive vibratile would then be well adapted to the later stage in which the spermatozoon has reached the tube, where its success in conjugation is dependent on its finding the ovum rather than on further progress.

"The stationary bunting type of motion is that which would be demanded by the passage of the spermatozoon through the egg membrane which has been so often observed in the lower animals. This very plausible explanation is, however, necessarily theoretical and must always remain so, as the conditions which surround the specimen on the field of the microscope vary in so many respects from those in which it accomplishes conjugation in the course of nature; but the practical importance of the study of types of motion is not affected by their explanation."

The duration of motility is a variable factor, dependent entirely upon the environment in which the spermatozoa are placed. Within the body they usually survive at least a week. One author describes a case in which he found living sperms in a woman who stated that coition had not been experienced for three and one-half weeks. It has been stated with regard to human sperms, that their motion should continue or be capable of being re-established for twelve hours. Cary (35) states, "first, that in their proper medium and at the body temperature the viability of the sperm cells may extend over a period of a few days; second, that their prolonged vitality is probably dependent upon the normal lime salts of the prostatic fluid; third, that the sustaining power of the seminal fluid is increased by its union with the normal secretion of the female genital tract." After death of the male animal they retain their motility in the genital tract for twenty-four to forty-eight hours.

Wolf (36) worked on this problem in rabbits, and summarizes as follows: "The motility of rabbit spermatozoa can be preserved for at least nine days by placing the juice of the epididymis in the Tyrode solution which had been buffered and to which glucose had been added. The solution is adjusted to a pH of about 7.4. Oxygen is passed in and a suitable amount of sodium bicarbonate added. The preparation must be kept at a temperature near the freezing point of water." Under ordinary conditions motility persists outside the body only a few hours after ejaculation, but if the semen is kept quite cool till the time of examination on a warm stage, motility should be capable of being restored in at least a fair percentage of the sperms for twenty-four to forty-eight hours. The cells are more sensitive to heat than to cold, and even to dilute acids more than to alkalies.

Henle (quoted by Ellenberger) states that a spermatozoon under favorable conditions travels at the rate of twenty-seven millimeters in seven and one-half minutes, which makes three and five-tenths millimeters per minute. This is about sixty times the entire length of the spermatozoon, and twenty-one centimeters in an hour. Forward motion is also more pronounced when the swim is against the current, such as is produced by the cilia of the oviduct. It has been demonstrated that feebly motile sperms become very actively motile when placed on the mucosa of a fresh Fallopian tube.

# Technique

The material used in the work came from abattoir animals, bull calves and adult bulls raised in the department herd, and sires upon which clinical observations had been made by various veterinarians in the field. Semen samples, many of which were sent in, were collected as often as possible after the method described by Williams (16). The genital organs were removed with as little chance of contamination as possible, and taken or sent to the laboratory where the examinations were made soon after arrival.

All cultures were made by searing the surface carefully, tearing out a small portion of the tissue with sterile forceps, and placing it upon the media. In most cases, however, where fluids were present, tubes were inoculated with the material which had been drawn off with a sterile pipette. As stated by Carpenter (9), in his work on the female genital tract, the organisms usually live in the depths of the tissue. The media used principally were glucose glycerin agar (glucose 1 per cent, glycerin 3 per cent); plain agar, both with a pH value of 7.4, and Loeffler's blood serum. Small amounts of sterile blood serum or defibrinated blood were added to most of the agar slants to insure better growths of streptococci when present. All tubes, to which the serum had been added, were incubated for forty-eight hours before inoculation to insure absolute sterility.

After inoculation, the agar tubes were sealed with sealing wax to give a partial oxygen tension which was quite necessary in isolating the streptococci. The growth of other organisms was by no means hindered by the procedure, for one tube from each organ was often left unsealed. Incubation was at 37° C, and the routine method of examining the tubes was identical with the method of Carpenter (9).

Whenever possible, a sample of blood was obtained from the animal either before, or at the time of slaughter, for agglutination with *Bact. abortum* antigen. Extracts from the seminal vesicles, testes, and epididymes were injected into the guinea pigs and examined at the end of four to six weeks for the presence of *Bact. abortum*.

Sections of all organs were fixed as soon as possible in either Zenker's or Helly's fluid. Hematoxylin and eosin were used as routine tissue stains. Eosin and methylene blue, and Mallory's connective tissue stain were, however, frequently utilized for special staining reactions.

The motility of the spermatozoa is best observed about half an hour after ejaculation, when the thick tenacious clot has started to liquefy. A drop of the fluid is placed upon a warmed slide, preferably one with a slight depression in it, and observation made with high or low powered objectives. The semen may be examined whole, or diluted with physiological saline solution. In the latter case, the sperms have a greater opportunity for freedom of motion in the absence of the thick viscid coagulate. A small vial of saline solution may be carried in one's pocket where it will be kept warm, and a drop of this placed upon the glass slide. If the clot of semen is merely touched to this drop on the slide, plenty of spermatozoa will be deposited for an examination. This method is very satisfactory for the observation of motility, but needless to say, the undiluted semen must be used for the determination of the number of sperms present. If necessary, the specimen may be covered with a cover glass and the oil immersion objective used. While a warmed slide is quite sufficient to enable one to detect the presence of motility, the field soon cools and the sperms gradually become less motile. If possible, it is best to use a small electrically heated stage warmer, which keeps the field at a constant body temperature, so that the duration of the motion may be observed for hours if warmed physiological saline solution is added as the fluid evaporates.

Stained preparations are best made with *thin* smears on the glass slides. This is conveniently done by placing a drop of the semen on a slide and smearing it over the surface with the edge of another slide. A fairly thin and even field is thus obtained. A still better method is to first dilute the semen with physiological saline solution, so as to obtain fewer sperms in the field. After drying the preparation in air, fixation may be produced by drawing the slide through a gas flame several times, by immersion in equal parts of alcohol and ether, or even by the use of tissue fixers such as Helly's or Zenker's fluids. For ordinary staining, heat fixation is the quickest, and at the same time is quite satisfactory. After the slide is cooled, or washed, to remove the fixing solutions, it should be placed for a few minutes in a *freshly* prepared solution (1 per cent) of chlorazene, as recommended by Williams, to remove the mucus and proteid material which otherwise blur the field. Other authors (38) have recommended diluting the semen with about twenty volumes of a 0.12 per cent sodium carbonate solution in 0.8 per cent sodium chloride. From this liquid the cells should be centrifuged for several minutes, removed with a pipette, and smeared on the slide. Following this, the slide should be thoroughly washed, preferably for ten minutes in running water, after which it is ready for staining. Numerous methods have been used for this procedure, but the sperms are more or less erratic in their reactions to the dyes, and one must be very careful to use the same method in all samples, in order to obtain uniform results. For quick staining, to bring out gross abnormalities of structure, and number of sperms present, one may use Gram's stain, or a light stain with any of the aniline dyes, such as fuchsin. To bring out the finer structure, particularly of the head, more careful technic must be employed.

Carnett and others (38) recommend the following: "The method of staining by iron-hematoxylin, particularly when supplemented by a cytoplasm stain, has proved, on the whole, the most satisfactory, and possesses the additional advantage of being absolutely permanent, a quality that few anilines can boast. The method consisted of treating the fixed object—and here the fixing agent was heat—with a two per cent solution of iron-alum for from two to four hours. The excess of iron-alum was then *completely* removed by pure water, and the object treated with a solution of hematoxylin (one per cent aqueous) for twelve hours or longer. The cells by this time were perfectly black. However, a 1 per cent solution of iron-alum removed the stain from the cytoplasm, leaving the chromatin of the head, the centrosome, and the axial filament a brilliant blue-black. Care must be taken that the preparation is not over-decolorized. After decolorization a saturated aqueous solution of eosin was added for from one to three minutes. This stained the protoplasmic envelope pink, and, unless the envelope is overstained, the view of the inner structures is not impaired in the least."

Williams (17) recommends using two staining solutions, one of alcoholic eosin and fuchsin, the other a diluted methylene blue. The results obtained are, however, more or less erratic, due to the unstable character of the former stain, and the ease with which one may over or under stain. Many beautiful specimens may, nevertheless, be obtained by this method. I have frequently used a fairly quick method, though one not satisfactory in all cases, which consists of staining for five or six minutes in a saturated aqueous solution of fuchsin, washing in water, and counterstaining for a few seconds in a strong solution of methylene blue. A quite satisfactory method is to stain from two to five minutes in a saturated aqueous solution of methyl green, with the application of gentle heat. The heat may be applied by warming the slide over a gas flame as it steams, or by placing the jar containing the stain in a hot water bath. The slide is then washed thoroughly and counterstained for five minutes in a strong aqueous solution of eosin. This is a fairly reliable method, and many excellent preparations may be obtained by its use. The nucleus is stained green, the anterior part of the head and all of the tail pink. So far, I have found this a very reliable stain for routine work.

# Pathology

In the genital tracts that I have studied, a complete pathological and bacteriological examination was made wherever possible, but in many of the abattoir animals, and certain others, gross and microscopical examinations only could be made. The genital organs of one hundred and ninety-six males have been examined, and the gross or microscopical changes, or both, determined. The abattoir animals were from a large slaughter house, and a small local plant.

Of the tracts, the pathology of which was studied, two were from aborted fetuses; seven from apparently normal young calves; four from mature fertile bulls; and sixteen from mature infertile or sterile sires. The remainder (167) of those examined were from abattoir animals. Besides these, three specimens of seminal vesicles, and seven of testes were studied.

The tracts of the aborted fetuses and veal calves were apparently normal, both on gross and microscopic examination of the vesicles and testes. On gross examination, the tracts of the mature fertile bulls were normal, except for the presence of many fine connective tissue tufts and strands upon the serous covering of the tails of the epididymes, and adjacent portion of the parietal layer of the tunica vaginalis. Microscopic sections of all parts were apparently normal. The more important pathological changes in the tracts of the sixteen sterile or infertile bulls are given in the appended chart. The tracts are numbered the same as in Group VI of the report of the bacteriological findings; that is, any particular number in either table refers to the same animal. References are made throughout the text to some cases which appear in this group of animals. Prostate and Cowper's glands are not included in the chart as they were not examined in some, and were negative in the others. Fibrous tufts and strands were present on the covering of the epididymes in each animal.

The study of sections from the abattoir animals, as well as those from the sterile or infertile bulls, forms the basis for the following observations upon the pathology of the male genital tract. The tracts secured from the abattoir were studied for the most part on the basis of the organ rather than on that of the animal. For example, all sections of testes were placed in the same bottle of fixer, and the same plan followed for the other organs.

TESTES: The testes seldom presented gross alterations of structure except for abscess formation, which, according to Williams, occurs more frequently in the bull than in any of the other domesticated animals. He also states that arrest in development by which the organs remain soft, flaccid, and somewhat smaller than normal is not uncommon. One very interesting specimen, which typifies abscess formation, came from a bull with a history revealing that one testis had become much enlarged, hot, and painful. These symptoms developed very rapidly. Anorexia was well marked. Local applications were used for several weeks, but at the end of two months the condition was so little improved that unilateral castration was performed. The general condition of the animal soon improved, but after a year of service he was so uncertain compared to what he had been, that he was sent to the butcher. It was impossible to obtain the other testicle for study, though it undoubtedly was abnormal. The testicle removed was considerably enlarged, measuring twenty by ten and one-half centimeters. The tunica albuginea presented a thickness of six millimeters, and was made up of firm sclerotic tissue. The epididymis was not recognizable in the mass. Testicular tissue was almost entirely gone. The only remains, of what appeared to have been parenchyma, was an elongated irregular area at one side of the organ. This tissue consisted of a whitish opalescent material, speckled with varying sized abscesses. This organ is pictured in Fig. 3. The remainder of the organ consisted of a thick yellowish caseous mass. *Streptococcus viridans* was recovered from the outer portion of the organ, and guinea pig inoculations failed to demonstrate *Bact. abortum*.

Microscopically, changes are quite common and varied in character. In the seminiferous tubules, the changes range from a slight desquamation of the germinal epithelium to atrophy and complete degeneration of the entire tubule, as was the case in the left testis of Bull 1. In the mild cases, spermatogenesis occurs apparently in a normal manner up to the spermatid stage, at which point many of the cells degenerate and slough off. These appear in the seminal fluid, associated with the few sperms that reach maturity. This sloughing and degeneration may be localized in a few of the tubules, or it may be widespread over the entire organ. Likewise, the changes may involve not only the more mature cells, but they may be so severe as to cause almost total degeneration and desquamation of the seminal epithelium, as in Fig. 15. These defects in spermatogenesis are of course evidenced in the semen by the presence of immature, or abnormal types of sperms. With cessation of spermatogenesis or degeneration of the epithelium of the entire gland, no sperms are formed. Not infrequently one finds numerous tubules, or even the entire testis in which the germinal epithelium is intact, but there is little or no evidence of mitosis, as in some tubules of Bull 6. The cells are several layers deep as in the normal condition, but they are not dividing. This condition is shown in Fig. 13.

In the more chronic forms, the tubules become atrophied, and frequently disappear entirely. The membrana propria may become thickened, due to excessive connective tissue formation, or infiltration with serum or exudate. On the other hand, a distinct atrophy may occur. The stroma of the organ not infrequently is thickened by inflammatory exudates, or by a noticeable increase in the connective tissue. In some testes, the connective tissue is so much increased that the tubules rapidly become atrophied, and disappear. In abscess formation, due to acute inflammations, the entire organ becomes enlarged, markedly hyperaemic, and infiltrated with leucocytes. Necrotic areas appear here and there in the parenchyma. The rete often shows a marked degeneration of the lining epithelium, and atrophy caused by increase of the interstitial connective tissue.

EPIDIDYMIS: This organ not infrequently presents gross abnormalities, and very often is pathological on microscopic examination. Acute, inflammation, with induration or abscess formation, is very common in the tail, but less so in the head and body. Possibly this is caused by the fact that the tail is the most pendant portion of the organ. In these cases, the tail is enlarged, soft, and quite hot and painful on physical examination. Enlargement, due to a connective tissue induration, occurs occasionally in all three parts, and the inflammation may produce adhesions to the adjacent serous membranes. Inflammation of both the parietal and visceral layers of the tunica vaginalis is very common. In those cases, the membrane usually is quite hyperaemic, and on the surface it presents many small reddened tufts of newly-formed connective tissue. In adult bulls it is exceptional not to find at least slight evidence of some previous inflammation. In all of the numerous bulls examined, both apparently normal and sterile, I have found but one in which some evidence of inflammation (either present or past) could not be found. Along with the fibrous tufts, are numerous fine strands of connective tissues passing from the covering of the tail of the epididymis to the adjacent portion of the parietal layer of the tunica. The strands often extend even to the upper part of the head.

Microscopically, inflammation of the part is shown by hyperaemia, loss of cilia of the lining cells, and exudation. In the more severe forms, the lining cells which furnish considerable secretion for the nourishment and stimulation of the sperms, become degenerated, and are exfoliated into the lumen, as in Fig. 21. This condition is very common in sterile bulls, and those of lowered fertility. In the chronic types, the interstitial connective tissue is increased in amount, leading to degeneration and atrophy of part or all of the tubules, as in the case of Bull 2. Infiltration with leucocytes, and necrosis, are the predominating lesions in the pyogenic types of inflammation.

DUCTUS DEFERENS: This tube seems to be peculiarly free from severe inflammatory processes, and when these appear they are limited to the mucosa. The cells of the lining membrane not infrequently show a mild type of degeneration and exfoliation, or in the more chronic forms, the entire membrane degenerates and disappears. In man, the duct occasionally becomes occluded, but so far I have failed to find this condition in the bull. Undoubtedly, when the occlusion does occur, it is near the origin of the duct at the tail of the epididymis.

SEMINAL VESICLES: The seminal vesicles and epididymis, especially the tail, seem to be the parts most subject to extensive pathological changes, and bacterial invasion. In most instances, diseased vesicles present gross manifestations recognizable on clinical examination, while on the other hand microscopic changes may be present in the absence of gross lesions. As diagnosed on physical examination, or even on post mortem examination of the tract, the various forms may be classified into:

1. *Acute Catarrhal Type*: In this form, the vesicles are usually enlarged, soft, and more or less reddened by hyperaemia. On physical examination of the animal, distinct flinching is produced when pressure is applied to the organ. Enlargement may even be absent in the early stages, and the diagnosis may be made from the extreme sensitiveness alone.

2. *Suppurative or Cystic Types*: In both of these types, the vesicles are usually enlarged, either uniformly, or, as is usually the case, in localized areas. The suppurative form may extend over the entire gland, forming one large encapsulated abscess, or on the other hand, it may take the form of variable sized abscesses with thick sclerotic, or thin fluctuating walls. Occasionally the abscesses rupture and discharge their contents into the rectum. Dr. Williams presented one case of this type. One vesicle was apparently normal, whereas the other was about five times larger than normal, and consisted of a dense outer capsule which was adherent to all surrounding parts. On dissection, it was found that the organ consisted of abscesses of various sizes, the larger one of which had ruptured some time previously into the rectum, leaving the distinct remains of an opening into that part. The cystic form may occur either with or without suppuration. One case came to my attention in which both vesicles were made up of abscesses of varying sizes as well as of a smaller number of cysts. Evidently the cysts were of the retention type, and were secondary to the pyogenic infection.

3. *Chronic or Sclerotic Type*: This form is characterized by a distinct firmness with or without marked enlargement. The condition may be accompanied by disease of the parenchymatous tissue or it may take the form of a chronic productive inflammation of the interstitial tissue. This inflammation may be simply a superficial thickening, or it may extend in between the lobules.

4. The *Peri-vesicular or "pan-inflammatory" Type* usually is the result of severe inflammation of the vesicles, with probable rupture of some of the smaller cysts or abscesses upon the surface. The vesicles are, as a rule, considerably enlarged and buried in a dense mass of adhesions which involve neighboring structures. The vesicles cannot be palpated on physical examination, and it is only on careful post mortem dissection that they may be studied. This type, however, is quite rare,—two cases only having come to my attention. In both, the vesicles themselves were markedly affected.

Microscopically, changes in the vesicles are quite frequently encountered, even in the absence of gross manifestations. In the acute catarrhal forms, the mucosa and submucosa are hyperaemic. The lining cells show various forms of degeneration, and there are, as a rule, inflammatory exudates in the lumen. As the inflammation progresses, the lining cells degenerate further, and become cast off into the lumen of the glandular cavities, as in Plate VI. The normal clear mucous secretion becomes mixed with fibrin, leucocytes, and cellular debris. These changes may involve merely parts of the organ, or they may be quite extensive. With large sections, one may find the inflammation in all stages, from the mildest catarrhal type, to complete degeneration and exfoliation of the secretion-forming mucosal cells, and filling of the cavities with degenerated cells, leucocytes, and debris. Frequently the interstitial tissue is in no way affected, but at times it is thickened by oedematous exudates, leucocytes, and fibrin. The chronic interstitial form is characterized by a considerable increase of connective tissue,—producing marked atrophy, or even complete obliteration of the glandular cavities. Microscopically the suppurative form may be diffuse over the entire gland, or as stated previously, may be in the form of localized abscesses, with or without a thick connective tissue wall. The parenchyma in these cases is usually extensively degenerated and atrophied in those parts that have not undergone suppuration and necrosis. The cysts appear to be of the ordinary retention type, and may or may not be accompanied by extensive changes in the lining epithelium.

Both the abscess formation and cystic conditions are undoubtedly initiated by an obstructive inflammation of all or part of the excretory duct. This is, however, a protective mechanism, for where the duct is closed the bacteria and exudates are unable to reach the urethra and contaminate the semen.

PROSTATE AND COWPER'S GLANDS: These glands were more or less neglected in the early part of the work, but later were subjected to the same examination as other parts. Of the thirty-six of each type of gland examined, I failed to find one with any gross changes, but two prostates were found that presented a mild catarrhal inflammation of the mucosa. It is probable that Cowper's glands, as well, occasionally undergo inflammatory changes.

SEMEN: The semen, made up as it is of mixed products of the testes and accessory sexual glands, is very often abnormal, as would be expected in view of the frequency with which changes occur in the glands contributing to its formation. The normal semen is remarkably adapted to its function of nourishment and stimulation of the spermatozoa, and their conveyance to the internal female genital organs. The spermatozoa are extremely sensitive to changes in their environment, with the result that any alteration of the physical or biochemical content of the seminal fluid may cause death of the sperms. With this in view, we must remember that disease of any of the contributing organs is a potential danger, and threatens the potency of the animal. Each or all of the glands may add bacteria, acid secretions, or inflammatory exudates. On the other hand, they may not function at all. In each case, however, the semen is altered.

Unfortunately it is impossible with present methods to obtain the fluid absolutely free from vaginal mucus, but with care it may be secured reasonably free from contamination by douching the prepuce of the bull and vagina of the cow before service. This method was used as often as possible in collecting the samples. The usual amount of semen obtained was from six to ten cubic centimeters.

With a hypersecretion of one or all of the glands, the semen becomes quite thin and watery, with a deficiency of solid matter, together with changes in reaction. On the other hand, hypofunction results in a secretion too viscid, which is equally unsuited to the requirements of the spermatozoa. The thin watery semen clots imperfectly or not at all, and clotting is essential in protecting the spermatozoa from the acid secretions of the vagina. Likewise, a medium too viscid is a distinct hindrance to motility. Changes in reaction are very frequently encountered. The sperms are very sensitive to dilute acids, so that with even a slight acidity motility may diminish or entirely cease. Purulent inflammatory exudates are occasionally mixed with the semen, and although the pus cells themselves have not been found to be destructive to the sperms, certain degeneration products in the exudate are very toxic, and inhibit or destroy the motility. So far, I have failed to find red corpuscles present. One very interesting sample of semen was quite thick, of a yellowish green color, and of a distinctly acid reaction. The secretion from the vesicles was later found to be of this same character, and was due to a *Ps. pyocyaneus* infection. The vesicles were highly inflamed and degenerated. The spermatozoa were in this case markedly decreased in number, and devoid of motility.

The early precipitation of the "Boettcherchen" crystals seems to be intimately connected with sterile semen, or spermatozoa of lowered vitality. Likewise, a decrease in solid matter is often seen in a deficient secretion. In normal semen, the clot disappears after standing a time, and a thick sediment settles out. This sediment is decreased in amount as a rule in abnormal semen.

SPERMATOZOA: Spermatozoa, the essential germinal elements, are very frequently abnormal, changes in which may be manifested in many ways. We may divide the deviations into changes in structure, and changes in the motility which is so indicative of the intrinsic vitality of the sperm. Reynolds (34) describes two forms of abnormal motion. The first is "rotary swimming," in which the sperms move forward progressively, and sometimes with fair rapidity, but in a spiral screwlike manner. He states that this type of swimming is very awkward, easy to recognize, and is usually of quite long duration. The other form termed "pendulum swimming," he states, is less common than the rotary swimming and is usually confined to relatively fewer sperms in a given field. "In this the middle piece and upper tail seem to lose their flexibility and balance to a considerable degree, and the lower tail motion swings the forward part of the spermatozoon to and fro with a pendulum movement. This type of swimming yields very poor progress."

One factor we must bear in mind in the study of the semen obtained from the vagina, is that the spermatozoa may be highly motile before ejaculation, but the admixture of hostile vaginal mucus may inhibit or destroy the motility. On the other hand, the conditions may be reversed. Cary (35), in one instance, found that the spermatozoa in a sample of semen collected from a condom, appeared to be of very low vitality, while when they were mixed with the vaginal secretions, an exaggerated activity was manifested. May we not have to contend with this factor in some herds in which there is a very distinct acid and toxic vaginal secretion from the products of cervicitis and vaginitis?

In a study of motility, we must consider not only the abnormal types which may be encountered but the percentage of motile cells, and the duration of the movement. In *necrospermia* all the ejaculated cells are motionless or dead. In other specimens, varying percentages of the cells are without motion, and the others may be possessed of full and lasting motility. On the other hand, the motility in some cases is very active at first, but quickly subsides even under the best of conditions. The appearance in freshly ejaculated semen of numerous sperms that have a tendency to take on the "undulatory tactile" type of motion when they should be in a highly active state, is very indicative of lowered vitality. Many specimens present this very picture, whereas the very active progressive movement should, under proper conditions, survive for a considerable time before it gives way to the second, and slower type. The cells frequently early bunt into epithelial cells or clumps of immotile sperms, then back out and move around sluggishly, only to repeat the same performance till they stop moving entirely. I have seen one specimen in which the sperms all tended to clump. Whether this was the result of some agglutinative substance in the vaginal secretion is problematical. I have seen several specimens of semen in which practically all the sperms were motile when first examined, but the motion did not survive for any great length of time. Even a small percentage of motionless sperms or of those showing lowered vitality is a considerable factor in potency. Although millions of the germinal elements are ejaculated into the vagina, large numbers of them are destroyed or become motionless there, and a small number is left behind in the cervix and uterus; so that even though but a single sperm is required for fertilization, the chances of impregnation are diminished in proportion to the number of dead or defective sperms.

*Aspermia*: Absence of spermatozoa in the semen is rarely encountered, and is probably due either to total cessation of spermatogenesis, or to an obstruction at some point in the system of excretory ducts. I have seen but one case of this character. The semen of this bull was greatly increased in amount, and of a thin watery consistency. Due to lack of cooperation on the part of the owners, the tract could not be obtained for study. *Oligospermia*, or a diminution of the number of spermatozoa, is quite common, and is undoubtedly associated with defective spermatogenesis, either as a result of poor mitosis of the seminal epithelium, or degeneration of the elements before maturity. This condition may vary from the finding of only occasional dead sperms in the field, to but a slight decrease in the usual number of normal sperms observed.

Abnormalities in morphology may be classified into immature types, and deformities or imperfect development of the head and tail. Defective spermatogenesis occurs so frequently that it is not surprising to find spermatozoa in various stages of development cast into the excretory ducts. The various stages passed through in the development, from spermatogonia to adult sperm, are numerous, and it therefore is to be expected that we should see in abnormal semen many different immature forms. No classification of the various types can be made, but a clearer understanding of them can best be obtained by a review of the process of spermatogenesis.

Spermatocytes and spermatids are seen in the more severe types of defective spermatogenesis, and are relatively uncommon, while the more mature forms that result from the transformations of spermatid to adult cell are very often seen. Some of these intermediate types are large oval cells without distinct nuclei and as a rule with poorly developed tails. Cells with no tails or distinct nuclei, those with protoplasmic appendages to the head or tail, and various other types, are occasionally encountered. Most of these are motionless and incapable of producing impregnation. Others are active, but survive a comparatively short time. According to Cary, the production of the immature cells is an effort on the part of the testes to supply an abnormal demand, and their presence indicates that the fertility of the semen is impaired.

The deformities, which may be divided into cephalic and caudal groups, are also the product of defective spermatogenesis, or they represent a degenerative process induced possibly by abnormalities of the fluid environment. It is rather difficult, however, to distinguish between deformities and immature types. The two most common cephalic deformities are what might be called macro and microcephalic forms. In the former, the head is enlarged to a greater or less extent, it is usually defective in staining qualities, and its outline is indistinct, due to degeneration of the covering membrane. This type is seen in Fig. 32. Also the shape of the head is usually abnormal, being either quite rounded, long and narrow, or short and very broad. Cells with protoplasmic appendages, though they are more properly an immature type, occasionally give the head a greater volume. Microcephalic sperms vary from those slightly smaller than normal to those in which the head is represented by a slight knob. In some cells, the head is small and round, in others, short and stubby, while another type is normal in outline but diminutive in size. These forms likewise are, as a rule, deficient in staining qualities, and are undoubtedly degeneration forms, occurring either as the result of faulty development, or degeneration subsequent to their formation. Cary believes they are degeneration types because in the majority of cells the tail is apparently fully formed, and in the normal process of evolution the tail is the last part of the cell to be exhibited. Double headed forms are quite rare, but they nevertheless appear at times. Their significance is difficult to explain. Another very frequent deformity of the head is a marked constriction at the posterior part so that it is the shape of a pear or top as in Fig. 27. In some, the head is otherwise normal in size, while in others it is much elongated, as in Fig. 28, or considerably atrophied. A constriction at the middle of the head, as in Fig. 29, is not uncommon. Both defects are undoubtedly the result of nuclear deficiency, as the nuclear part of the head in these cases is much diminished in size, and stains very deeply or not at all. I have seen spermatozoa, the heads of which were like an inverted cone, with a bulging rounded base. Other heads are even somewhat contorted and bent on themselves, as shown to some extent in Fig. 26.

Under caudal deformities, the most frequent form encountered is a thickening of the connecting piece. This may occur as a uniform thickening, or as a bulging appendage. Rudimentary development of the tail, the presence of two poorly formed tails, and defective development of the connecting piece occur rather infrequently.

All these immature and defective types are, as a rule, motionless, and of course incapable of producing fertilization. Their presence indicates lowered fertility of the semen. Besides these deformities, there are sperms showing a curvature of the tail at an acute angle just posterior to the neck,—the so-called "wry neck." Their significance is difficult to explain, but they occur frequently in semen fixed and stained by the same routine methods used on samples in which they are absent. They probably are not the result of the methods used in fixing and staining. Some think they are slightly immature types, or that the condition is produced by abnormal contractions of the tail. The majority of sperms, however, especially those from highly fertile bulls, do not show this type at all.

The most common changes in the spermatozoa, are those in which there is a separation of the head from the tail, and degeneration of the head as evidenced by reaction to stains. The separation of the head from the tail always occurs at the neck, and often is associated with degeneration or abnormalities of the head. The separation, in the majority of cases, indicates some lowering of vitality in the elements, although in many instances traumatism produced in making smears or collecting the samples is responsible. Various forms of abnormal staining of the head are very common. The cell membrane, which is normally distinct and sharp, becomes blurred in outline. Normally, the head takes a good differential stain, the anterior part staining lightly, and the posterior part somewhat deeper. The nucleus is distinct in outline and well defined. The lighter "inner body" stands out in well stained specimens. As the result of degeneration, the whole head may take the stain uniformly, either slightly or much deeper than normal, according to the degree of degeneration. The whole problem of staining, however, depends very much upon the methods used, and the care with which they are applied. When a good method is obtained, it should be adhered to, and used uniformly on all specimens. As a rule, however, a certain amount of practice will enable one to differentiate between the sharply outlined, clearly staining normal forms, and those that show abnormal reactions to the stains.

# Bacteriology

A complete bacteriological study was made of the genital tracts of fourteen normal young veal calves (six to twelve weeks old), four mature fertile bulls, and sixteen mature bulls, either sterile or impotent to some degree. Together with these, the tracts of eleven aborted fetuses, seven calves dying from calf infections (scours or pneumonia), and sixteen bulls slaughtered at an abattoir were studied bacteriologically. Occasionally, studies were made of individual seminal vesicles or testes, when these parts alone were brought or sent in. The history of the abattoir animals was, of course, quite indefinite or entirely negative. On the killing floor, many tracts could be studied for pathological changes, but in the bacteriological work it was difficult to care for more than two tracts on each visit.

The results of the bacteriological examinations are given in the appended tables, in which the tracts are divided into six groups. The results in Group I. consisting of normal veal calves, indicate that the genital organs of young male calves are, under normal conditions, free from bacteria. Carpenter (9) obtained like results in examining the genital tracts of heifer calves. The cultures made from the seminal vesicles and testes of all these veal calves were, with two exceptions, negative. Both seminal vesicles of one tract and one of another yielded cultures of *Staphylococcus albus*.

Adult bulls of known fertility were naturally difficult to obtain, only the four animals in Group II being available for examination. Two of these (Nos. 1 and 2) were from the experimental herd kept by the department, and at all times had a good breeding history. The other two were good breeders, but were slaughtered because of poor pedigrees. Bull 1, raised in the department herd, had a severe attack of scours when a few weeks old, while the calfhood history of the other is not known, he having been purchased after reaching sexual maturity. The cultures from the genital organs of the former (Bull 1) were entirely negative, except those from the left epididymis and scrotal sac, which yielded growths of *Streptococcus viridans*. All the organs of the tract from this animal were normal, except for the fact that numerous strands of connective tissue extended from the serous covering of the tail of both epididymes to the adjacent part of the parietal layer of the tunica. The tract of the other failed to show any organisms. The only evidence of any abnormality was the presence of the same connective tissue strands on the tail of the epididymis, as in the first tract. The other two bulls gave negative cultures from all parts.

Of the sixteen bulls in Group III, slaughtered at abattoirs, and in which no history was available, eight failed to show the presence of any organisms in their genitalia. Of the others, the vesicles yielded cultures of *Staphylococcus albus* nine times, and *streptococci* four times. *Staphylococcus albus* was recovered once from the prostate, and once from Cowper's glands. The testes gave cultures of staphylococci in two cases, and *Bact. abortum* in one. No observable anatomical changes accompanied the presence of the Bang bacillus in this case. The epididymes showed growths of staphylococci five times, and streptococci on three occasions. *Streptococci* were isolated from the scrotal sacs of eight testes.

The results in Group IV (aborted fetuses) show that bacteria are often present in the seminal vesicles or testes of these animals. As a rule, however, the organisms are identical with those isolated from the blood or other parts of the animal. This is to be expected, however, for because of the feeble resistance of the fetus to any infection, the organisms circulating in the blood may be isolated, as a rule, from many different organs and tissues. All samples of blood set with *Bact. abortum* antigen were negative, irrespective of whether or not the organism was recovered from the blood or other tissues. This is in accordance with the findings of Carpenter in the female fetus,—the resistance is so feeble that few or no antibodies are formed to combat any existing infection. *Bact. abortum* was recovered in two cases from the vesicles, and in four cases from the testes, but in each instance the same organism was present in the blood or other tissues of the body.

The results from the tracts of the calves dying of calf infections are given in Group V, and show that five were negative. The other two showed *B. coli*, staphylococci, and streptococci, in the organs indicated by the chart.

In Group VI, the mature infertile or sterile bulls, there was a comparatively wide variation in the type of organisms encountered, but streptococci and micrococci were the most common invaders. In the order of the frequency of infection, the organs would be enumerated as follows: Vesicles, epididymis (usually the tail), scrotal sac, testes, prostate, and Cowper's glands. The first three parts mentioned usually contained bacteria. A streptococcus was the usual invader of the scrotal sac, and very probably was the cause of the connective tissue tufts and strands so frequently seen. The vesicles and epididymes gave, in the order of the frequency of their occurrence, staphylococci, streptococci, *B. coli*, and *Ps. pyocyaneus*. The streptococci were usually of the viridans group, though a few were hemolytic, and two strains were indifferent to blood. The testes gave growths in only eleven instances,—staphylococci eight times, streptococci two times, and an unidentified rod once. The prostate yielded staphylococci twice and Cowper's gland once.

As emphasized previously, the vesicles and tail of the epididymis are most subject to infection and degenerative changes. At the same time, they are intimately connected with the secretion of the semen. Once the epididymis becomes infected, there is nothing to prevent the organisms, together with inflammatory products, from being mixed with the semen and ejaculated during coitus. Also in the vesicles, unless the inflammation is so severe as to occlude the excretory duct, the organisms are mixed with the vesicular secretion, which is emptied into the urethra during ejaculation. Carried along with the bacteria, are, of course, toxic products, degenerated cells, and the otherwise altered secretion of the glands. One interesting case noted was that of a bull that had passed from a state of fertility to that of complete sterility during a period of two months. The semen was semi-fluid, greenish yellow in color, and contained a very few non-motile spermatozoa. Post mortem examination showed that the vesicles had undergone abscess formation and that they contained yellowish green material similar to that which had been discharged during copulation. *Streptococcus hemolyticus* and *Ps. pyocyaneus* were isolated from both vesicles, and from the semen. *Micrococcus albus* was isolated in nearly all cases of vesiculitis and was often associated with *Streptococcus viridans* or *hemolyticus*.

Bacteriological studies of the semen are, on the whole, more or less unsatisfactory, due to the present difficulty in obtaining samples free from any chance of contamination. In most of the abnormal bulls, bacteria of various types were isolated from the semen, most of which agreed culturally with those later isolated from the internal genital organs of the same tracts. The method of culturing consisted of douching the prepuce of the bull and vagina of the female with sterile saline solution before breeding. Samples of vaginal mucus were taken before service, and the flora compared to that after douching. This method of douching produced vaginal samples relatively free from bacteria, at least so much so that the post coital fluid demonstrated that many organisms must have been introduced from without. Whether or not they came in with the semen is problematical, but in all probability this was the method of introduction.

I have so far failed to obtain *Bact. abortum* from the tract of an adult animal, either by direct culture or guinea pig inoculation, except from the testicle of one abattoir bull. The agglutination tests with *Bact. abortum* antigen were all negative, except for two abattoir bulls. The results so far obtained would seem to indicate that, in accordance with the findings of other workers, the Bang organism seldom invades the male genital tract, or does not thrive there after its introduction. Schroeder (12) and others, have, however, on various occasions, recovered the organism from the bull, and the former author even states that it invades the vesicles and is eliminated with the semen.

# Discussion

A complete discussion of those factors which have a bearing on reproduction and fertility in an animal, includes not only a thorough study of the genital tract, but an appreciative consideration of various extrinsic factors. The effect of environment has long been known to have a marked influence upon breeding, particularly with reference to animals in domestication. Diet, though long relegated to a minor phase of the question, has, within recent years, come to be a matter of prime importance with regard to its bearing upon the entire body metabolism. The endocrine organs preside over and regulate the growth and functioning of the genital organs from the earliest embryonic stage to the cessation of sexual life. Any derangement in one results in functional or organic changes in the other. In a given mating, we must take into consideration such factors as impediments to coitus, as well as those numerous agencies in the female which may interfere with the union of sperm and ovum, or with the successful implantation of the fertilized egg in the uterus, and its growth and development there till normal parturition takes place. Successful reproduction depends upon the mating of sexually sound females to equally sound males. Considering the various factors which govern reproduction, sexual soundness must necessarily depend, to a large extent, upon a good general condition of the entire body.

ENVIRONMENT: The effect of environment on fertility in the bull is no doubt a minor factor. Cases in which changes in environment affect fertility probably occur, however, particularly when fear and other psychic disturbances play a part. Marshall (29) states: "It would seem probable that failure to breed among animals in a strange environment is due not, as has been suggested, to any toxic influence on the organs of generation, but to the same causes as those which restrict breeding in a state of nature to certain particular seasons, and that the sexual instinct can only be called into play in response to certain stimuli,—the existence of which depends to a large extent upon appropriate seasonal and climatic changes."

DIET: Under this heading we may include not only the effect of deficient food, but also constitutional disorders, as a result of which the organs of generation and those glands guarding their function receive insufficient nourishment. It is a well known fact, and long has been, that animals fail to breed when they are in a run down condition or when they are fed a deficient diet. Cary (35), quoting Hagner, states that the virility of the spermatozoa is often in direct proportion to the general physical condition of the patient.

Reynolds (34) emphasizes the fact that it is an established principle among animal breeders that a high protein diet in both sexes is essential to full fertility. "Oligospermia with deficient vitality of the spermatozoa is not infrequently found from constitutional disorders. It can easily be demonstrated in animals that both low diet and conditions of life that produce a nervous excitable state are attended by oligospermia." Animals that are closely confined, those that are over-fat (show animals), as well as those fed a deficient ration very frequently fail to breed, but exercise and change of diet soon overcome the impotency.

Dutscher, Hart, Steenbock, and other biological chemists have done extensive work to show the essential importance of vitamines and minerals in the diet. Their results indicate that animals cannot thrive and breed normally when fed a diet composed solely of the products of one plant. There must be variety, and there must be not only a correct nutritive ratio, but the mineral and vitamine content must be present as well. Cows fed on the products of one plant often failed to breed, and if conception occurred, it invariably resulted in a premature birth, or the birth of weak and poorly nourished calves. The work of these authors is fundamental, and brings out many important points. Is it not probable that the deficient diet results in weakened tissues which are easier prey to the invasion of bacteria?

Macomber and Reynolds (39) experimented upon white rats to determine the effect of defective diet as a cause of sterility. They call attention to the confusion caused by the application of the term sterility to most, or all, infertile matings. They believe that failure of reproduction is, in fact, the result of decreased fertility rather than of actual sterility on the part of the two individuals concerned. "There are certainly a large number of infertile matings which are purely functional and due to physiologic alterations or local conditions. Such physiological alterations moreover coexist in the sterilities of pathologic origin and when unrecognized and consequently unremedied, undoubtedly explain a large proportion of the continued infertilities after operation." Is it possible for a bull to be infertile to the cows in his herd that have been fed a deficient diet, and at the same time to breed well when mated to animals outside this herd? This is rather improbable in practice, but there is always the possibility of its occurrence. In the experimental work, white rats were used: one strain from the Wistar Institute with a fertility of about 65 per cent, and the other from a Dr. Castle's strain with a fertility of about 90 per cent. The Wistar rats were fed in groups, each group receiving a diet deficient in a certain substance: calcium, protein, or fat soluble vitamine. To this group was added a diet deficient in both calcium and protein (war diet). These diets reduced the fertility of the groups from the original 65 per cent, to 55, 31, and 14 per cent respectively. It delayed the appearance of fertility in young rats raised on these diets, and lowered its degree in the mature animals. Most of these rats, however, though infertile to each other, bred promptly when mated to the Castle rats of known fertility. This demonstrates clearly that relative infertility of given matings does occur. One interesting feature of the work is the fact that in the matings on the single deficiency diets, four deliveries of macerated fetuses occurred and there were two more in eight deliveries from those reared on the war diet. No cases of this kind had previously occurred in this strain, which had been under observation for several years. Does this throw any light upon the cause of macerated fetuses in cattle? Microscopically the testes and ovaries of these infertile rats showed no observable changes, a fact which is of great importance to bear in mind.

Williams, in his book on disease of the genital organs, brings out quite clearly the relation of defective diet, overfeeding, and lack of exercise, to reproductive efficiency.

Novarro (40) observed that pigeons fed on a diet without vitamine B showed degeneration of the seminal epithelium, with hypertrophy and hyperplasia of the interstitial cells of the testis. Another author (Allen) showed that reduction in the quantity of water-soluble vitamine in the diet of rats resulted in total degeneration of all the germ cells, but it did not interfere with growth and development in other respects.

The observations of Williams (41), in a pure bred beef herd in Hawaii, clearly demonstrate the intimate correlation between poor fodder as the result of extreme drought, and the accentuation of, or increased susceptibility to, genital infections, as demonstrated by clinical findings. The genital disorders started soon after the onset of the drought, and immediately took a downward trend with the advent of the rainy season.

Judging by the work quoted, we will observe that deficient diet, though it does not always affect the general health, has a profound effect upon the genital organs of both sexes, associated with disturbances of spermatogenesis in the male. In most debilitated and weakened conditions of the male, spermatogenesis ceases or is markedly defective. We must, undoubtedly, explain this fact upon the ground of deficient nourishment to the reproductive organs or possibly the endocrines. The vitamines have been termed nuclear nourishers, and their absence probably results in nuclear deficiency.

ENDOCRINES: Bell (42) emphasizes the fact that not only the structure but also the function of every part of the body is in close correlation with the rest. "This is essentially true of the ductless glands: the shadow of their influence is over all." Further he states that when we remember that the individual exists to perpetuate the species, it is not difficult to realize that the metabolic factors concerned in reproduction are the same as those related to the individual metabolism. It follows, therefore, that the ductless glands which regulate the individual metabolism concern equally the reproductive. Brown (43), discussing the same subject brings out the generalization that the sympathetic, since it is the most primitive part of the nervous system, is closely associated with the endocrine system, a still more elemental means of communication in the body. Also since specialized reproductive cells appear before the nervous system, the organs of reproduction remain closely associated with the older chemical reactions now specialized in the endocrine glands. "The endocrine glands, the reproductive organs, and the sympathetic nervous system, therefore, remain as a basic tripod, and it is not likely that a disturbance will occur for long in one limb without affecting the other two." The former author believes that the gonad keeps the other ductless glands informed of the needs of the genital tract, they in turn influencing the general metabolism. Jump (44) states: "We are therefore justified in believing that there is a correlation of function between these (endocrine) glands and that some cases of sterility are probably due to a derangement of this correlation." Biedl (45) concludes: "There appears to be an intimate anatomical and physiological interrelationship between the different blood glands which is manifested clinically by the fact that the pathological disturbance of one gland is accompanied by symptoms pointing to the functional derangement of one or more of the others. Knowing, as we do, the many sided interactivity which subsists between the different internal secretory organs, it is readily conceivable that isolated diseases of single organs of this group are very much rarer than, at the first glance, they would appear to be. In the present state of our knowledge, the only course of investigation which is open to us is to start with the known results of the functional derangement of any organ, and, by following these up, to seek the primary link in the pathological chain."

Most workers seem to agree that a special connection exists between the normal function of the adrenal cortex and the sexual organs. Tumors of the former are usually associated with sex abnormalities, and feeding young animals the gland substance seems to stimulate growth of the testes.

Many arguments have been brought forward to show that the prostate produces an internal secretion. It is a well known fact that this organ atrophies after castration, and enlarges as the sex life dwindles. As has been previously stated, Serrlach and Pares reached the conclusion that the gland produces an internal secretion which controls the testicular functions and regulates the process of ejaculation. Also they state that if the prostate is removed, spermatozoa are no longer produced in the testes, and that the secretory activity of the accessory genital gland ceases. The secretion is, at any rate, a stimulus to the internal secretion of the testis.

The thyroid bears a distinct biological relationship to the sexual glands. Removal of the gland results in imperfect development of the gonads, infantilism, and general torpor. Bell (42) believes that the association between the thyroid gland and the genitalia is as intimate as the relation of the pituitary to the genital functions.

Of all the endocrines, perhaps the anterior lobe of the hypophysis is in most intimate correlation with reproduction. Castration results in hypertrophy of this organ, while removal of the anterior lobe usually leads to death. In those cases in which death does not ensue, it results in genital atrophy, stunting, and reduction of sexual activity. In young animals, spermatogenesis ceases entirely even after partial extirpation of the anterior lobe. Biedl (45) states that "in disease of the hypophysis, derangement of sexual activity occurs very early in the course of the disease, shown in women by the cessation of menstruation, and in men by impotence."

The thymus, as is well known, is quite intimately associated with the development of the genital organs. Its normal disappearance is always associated with the development of sexual maturity in the individual. Hewer (46) conducted experiments to ascertain the effect of thymus feeding on the activity of the reproductive organs in the rat. She concludes in part: "Male rats appear more susceptible to the influence of thymus feeding than female rats. With moderate doses of thymus, sexual maturity in the animals treated is delayed, a phenomenon which is attributed to delayed development of the testis. With large doses of thymus, in the male, the testis is structurally affected: in the young animal in the direction of retardation of development, in the mature animal in the direction of degeneration. This degeneration is confined to the testes. In the degenerating testis, cells of Sertoli appear to be absent: the spermatogonia are present, also dividing, and may lie free in the lumen of the tubule; spermatids, many with abnormal nuclei, are shed into the lumen in large numbers; spermatozoa are practically absent. In the later stages, only a few dividing spermatogonia appear among the debris of the other unrecognizable cells of the tubule. In the epididymis which itself is normal, when the testis is showing degeneration, very few spermatozoa appear, in the later stages none. Many spermatids are present in various stages, and some spermatocytes. Animals in the hyper-thymic condition appear to be sterile."

The foregoing references will, I hope, serve to bring out the facts that environment and diet, together with the general body metabolism and the endocrines, have a more or less profound effect upon the development, growth, and functioning of the genital system. In the experiments it has been shown that sterility is not necessarily accompanied by any apparent microscopical changes in the gonads, or even at times in the general body health. Nor can we exclude impotency of the male entirely even when the spermatozoa are normal in shape, and motility. Carnett, and others (38), years ago stated: "Indeed, there is abundant clinical proof to the effect that systemic conditions which have no appreciable effect upon the motility or conformation of the spermatozoa materially interfere with reproductive power." The entire complex genital system is inseparably linked up with the body as a whole, a fact which we must bear in mind at all times.

*Impediments to coitus* may be due to great difference in the size of the two mated individuals, psychic disturbances, or inability to protrude the penis. Williams (17) mentions several physical impediments, as deformity of the limbs or feet, sore feet, overloading of the rumen, obesity, fear of falling, and paralysis. Coitus may be somewhat delayed, or even not performed as the result of a severe inflammation with sensitiveness of the penis or prepuce. Occasionally tumors of the penis are encountered which may interfere with protrusion of the penis, or its entrance into the vagina. Not infrequently the penis is rendered incapable of protrusion as the result of inflammatory adhesions, tuberculosis of the preputial lymph glands, etc.

*Excessive sexual use*, within certain limits, probably has not, in itself, any material permanent effect upon the reproductive capacity. The frequency with which bulls used to excess break down sexually, is probably due to the devitalizing effect upon the tissues of the genital organs, this opening the way to bacterial invasion and other destructive influences. Over-use is probably not dangerous, unless continued over long periods, but at the same time it offers greater opportunity for infection to be introduced into the body from intercourse with large numbers of females. Lloyd-Jones and Hays (47) carried on very interesting experiments on the influence of excessive sexual activity of male rabbits on the properties of the semen. Their plan was to mate male rabbits in quick succession, and study the character of the semen on the first service, and every fifth service thereafter. The safe limit was twenty services in three hours. As would be expected, the volume of the semen, after the first few services, became gradually reduced in amount. "In rapidly successive services, the semen becomes less viscous and tends to lose its characteristic milky appearance until at the twentieth service, when the fluid is thin and watery." It seemed as though there was a well marked reduction in the number of spermatozoa per cubic centimeter in the advanced services. Successive copulations also resulted in a marked decrease in the number of motile spermatozoa, together with a shorter duration of perceptible vitality. The certainty of producing impregnation at the same time became less and less. "This reduction in the per cent of effective matings when the male is sexually overworked is recognized by those engaged in animal breeding as one of the most noticeable and universal concomitants of heavy sexual service."

In another paper, these same authors studied the effect of sexual excess upon the character of the offspring. In part, they conclude: "By no means thus far used has any inferiority of progeny from the heavy sexual service been discovered. They are fully equal if not superior to progeny from very light service of male."

*Infection* is without doubt the greatest single factor capable of producing functional and anatomic changes resulting in varying degrees of impotency and sterility. The changes produced range from the addition of the toxic products of bacterial growth to the seminal fluid, to the complete destruction of the parenchymatous tissue of one or more of the contributing sexual glands. Anatomic changes are by no means essential to the production of lowered fertility. As has been previously stated, the work on veal calves indicates that the genital organs of young bulls are normally free from bacteria. Likewise in normal adult animals, the bacterial content of the genital organs is as a rule low or negative. It is possible that a certain flora is normal for the tract at sexual maturity, as in several other organs of the body, but under the strain of sexual excess, defective diet, or other weakening influences, these organisms may become pathogenic. Streptococci and staphylococci have at times been found in apparently normal parts of the body, and at other times they are found associated with severe pathological lesions in the genital tract. The degree of pathogenicity is of course difficult to determine, except as we find the bacteria associated with abnormal conditions. Carpenter (9), however, injected streptococci into the genital tracts of female calves and produced lesions resembling very closely those from which the organisms had been isolated in adult sterile animals. Personally, I am inclined to believe that the genital organs normally are free from bacteria, or if any are there they are better able to multiply under the strain of devitalization of the tissues. Bacterial invasion, however, does take place quite frequently, but the paths of entrance of the organisms are somewhat problematical. Hematogenous origin is always possible, though it is rather difficult to definitely implicate this mode of entrance. The urethra is perhaps the easiest and most common path for the entrance of bacteria, though even here it is not possible to make definite assertions. Contiguous spread of infection from neighboring structures is very probable in some cases, particularly in pelvic peritonitis. The bacteriological results hardly bear out the theory of Williams that the organisms lie dormant in the genitalia of the animals until the advent of sexual maturity, at which time they acquire pathogenic powers. On the other hand, his clinical observations seem to indicate that this may be possible. Calves suffering from "calf infections" frequently do harbor organisms in their genital organs, but whether or not they persist there till sexual maturity is a matter of conjecture. The most logical theory seems to be that animals from herds in which genital infections are very severe, or those that have had severe attacks of scours or pneumonia, are more susceptible to those infections, due to the early lowering of their vitality. One bull in the department herd certainly had a severe ordeal as a calf, but as a mature bull he was highly fertile. Moderate sexual use and proper sexual hygiene probably had much to do with this. In the bull, infection of some part of the genitals, during some period of life, is very constant, however, whether or not it is productive of observable changes in his breeding efficiency. The finding of the fine connective tissue strands and tufts on the serous surface of the tail of the epididymis of practically all bulls examined, both sterile and fertile, indicates past or present infection of the scrotal sac. The vesicles and tail of the epididymis are, as stated previously, the most commonly invaded tissues of the tract. The testes are less frequently involved.

While it is difficult to obtain irreproachable proof that the bull is a disseminator of genital infections, the findings of clinicians quite clearly indicate that this is true, and laboratory methods tend to support this assumption. Williams believes that not only may the bull infect the female with organisms which interfere with the given conception, but that he often implants there organisms which interfere with future pregnancies, and even with the life of the individual in some cases. The high abortion and sterility rate following the use of certain sires, and the appearance of characteristic infections after service to certain bulls, clearly indicate that in all probability the bull does eliminate with his semen those organisms which produce lesions in his genital organs, and are capable of infecting the female. W. L. Williams (48) cites the case of a pure bred herd in which breeding had progressed satisfactorily until heifers had grown to breeding age and a second bull was obtained. "Some cows of the old herd were also assigned to the young bull which had not previously been in service. The cows bred to the old herd bull continued to breed normally. The cows and heifer's bred to the new bull conceived with difficulty or not at all. Those which conceived mostly aborted, and those which calved had metritis and retained fetal membranes. The two first cows in which pregnancy terminated died of metritis." I have frequently had semen samples sent in from bulls that were not only failing to get cows with calf, but following each service the females showed a severe vaginitis. W. W. Williams worked in a herd in which service to certain bulls was in each case followed by a severe vaginitis and cervicitis, only to be followed later by a characteristic salpingitis.

Vaginal smears taken before and after service, in many instances, show that in all probability bacteria, especially the streptococci, were deposited there with the semen. These results have been obtained upon several occasions, at which time the vagina was usually douched prior to each service with sterile saline solution. Streptococci and other organisms have been isolated from the vaginal samples obtained by this method. In most cases, they were absent from samples taken before service. Extraneous contamination, and error, must be taken into consideration, but the results tend to bear out clinical observations that the bull is probably a disseminator of some infections associated with the genital organs of both sexes. At any rate, organisms have been isolated repeatedly from the genital organs of the bull, of the same biological character as those which are associated with sterility, abortion, and allied phenomena in the female. In the absence of obstruction in any part of the tract, there is nothing to hinder infection from gaining access to the seminal fluid, and being excreted during ejaculation.

Hopper (18) states: "A diseased bull may manifest non-fertility or decreased potency in different ways—by repeated service to apparently normal females without conception, by a high abortion rate in females that have been apparently normal, by characteristic infections following the use of any particular sire, or by abnormalities in the breeding tract noted by rectal or physical palpation."

Admittedly, *Bact. abortum* has little affinity for the genitalia of the bull, though Schroeder states that the bull harbors the organisms in his seminal vesicles and that they are eliminated with the semen. Other authors have occasionally isolated the organism from the vesicles, testes, or both. Schroeder's theory that infection of the female occurs indirectly by contamination of the fodder with the semen is probably rare in occurrence. The very limited number of cases in which investigators have demonstrated the presence of the organism in the male genital organs, and the apparent immunity of the bull to the bacterium as determined by the agglutination reaction, seem to indicate that he plays a small part in the spread of this type of infection in the herd. On the other hand, it would seem that he is at times intimately associated with the spread of certain other organisms that interfere seriously with herd reproductivity.

The diagnosis of infertility and sterility rests upon a thorough physical examination of the genital organs, together with a detailed study of the semen. The history of the animal and herd involved must also be very carefully inquired into, especially the part covering the result of every service by the sire in question. Besides this, we must always consider all factors which have a bearing upon the subject, remembering the physiology of each part, and the role it plays in reproduction. Bacteria gain entrance to many parts of the tract, where they multiply and probably add toxic products to the seminal fluid, altering its biological character and resulting in partial or total destruction of the secretory tissues. The testes, epididymes, vesicles, prostate, and other parts, each contribute their essential part to the semen, abnormalities of any one of which, as a rule, result in interferences with reproduction. If the vesicles are involved, we must bear in mind just what is the part played by their secretion, and what is the probable result if their essential elements are not added to the semen. In like manner, we must consider the prostate, whose secretion stimulates the vitality of the sperms, and adds fluid bulk to the semen. Extirpation of the vesicles or prostate alone results in lowered fertility, without altering the sexual desire, while removing both glands produces total sterility. Partial or total destruction of the parenchyma of either gland produces the same effect as extirpation, in that its function is altered or entirely absent.

The semen should be examined, not only for the number of spermatozoa and the percentage of those that are motile, but for the duration and type of motion. Normal semen, when first examined under the microscope, shows a field closely packed with highly motile spermatozoa. In every study of the semen, however, we must bear in mind the temperature and other conditions under which it has been kept since emission. On the other hand, semen from bulls of lowered fertility shows changes ranging from mild disturbances such as sluggish motility and a slight decrease in the number of sperms present, to aspermia, or total lack of motion. Normal semen, when compared with abnormal specimens, as a rule presents distinct differences, either in motility, staining properties, or structure of the spermatozoa. Impotent bulls, however, may show at times few or no observable changes in their genital organs. The only assumption here is that the condition probably is of endocrine origin, or is some functional disturbance. Of oligospermia Reynolds states: "Oligospermia, with normal motility and vitality, is not absolute sterility, but is of high importance because the percentage of destruction of spermatozoa during their passage through the genital canal of the female is so enormous that the possibility of impregnation by semen which starts out with a deficient number is always poor. When the genitals of the female partner are in a condition which is even moderately hostile to the spermatozoa, impregnation by such semen becomes so unlikely as to be not even a probability."

Motility may be lacking in a small number, its absence may be observed in a large percentage, or even in all those in the field, as in necrospermia. On the other hand, the motility may be sluggish or of abnormal types in variable percentages. Sperms with sluggish motility are always low in vitality, and have weak powers of insemination, as the motion lasts but a comparatively short time. The vitality may be but moderately lowered, so that although the sperms are highly motile when ejaculated they soon lose their power of propulsion. The type of motion is likewise an indicator of lack of vitality. The "progressive vibratile" motion described by Reynolds should proceed to a high degree for a long period before the "undulatory tactile" or bunting types of motion appear. Early appearance of these two latter types indicates in most cases a marked lack of vitality of the elements. The motion should be vigorous and lasting, for, as stated by Reynolds, "nothing is more certain than that spermatozoa of merely moderate vitality seldom impregnate a female."

The early precipitation out of the "Boettcherchen" crystals is very characteristic of oligospermia, and impotent semen. The theory here is that crystals do not precipitate out when a fluid is actively moving, but soon do so when the fluid is motionless. The sediment which normally makes up about two-thirds of the sample is usually decreased in abnormal samples. The semen itself should be observed for unusual viscosity or a thin watery condition. Clotting should occur readily after emission, but the clot soon liquefies to some extent, allowing the spermatozoa to become more active. This clotting is, of course, to protect the delicate sperms from the hostile secretions of the vagina.

The presence of immature and deformed types of sperms represents some disturbance of spermatogenesis, but it is difficult to explain the significance of these forms. They are seldom seen in normal samples, and undoubtedly none are capable of producing impregnation. In the case of minor abnormalities of staining reactions, the sperms are probably deficient in nuclear material or otherwise altered so that probably they are incapable of reaching and uniting with the ovum. When impregnation does occur in these cases, weak offspring undoubtedly result in many instances.

The work has by no means progressed to the point where one may, by an examination of the semen, determine the degree of impotency with great accuracy, or even whether the animal may be restored to sexual health by proper hygienic and therapeutic treatment. Relatively, the greater the changes in the semen and spermatozoa, the less the chances of impregnation. Infertility to any marked degree, is, however, usually accompanied by corresponding changes in the seminal fluid and its germinal elements.

Examination of the semen is, and probably always will be, simply an aid in reaching a diagnosis. While abnormalities of the semen and spermatozoa are associated with sterility or infertility, it is unwise to lay too much emphasis upon this method of diagnosis alone, especially with regard to the making of a definite prognosis. When large numbers of abnormal spermatozoa are present in the semen, we are safe in saying that the animal is, at the time, of lowered degree of fertility. One should be very cautious, however, in foretelling how long the condition will last, or if the animal may in time be restored to full fertility. Sterility, due to organic disturbances, probably seldom yields to treatment, but when it is due to functional disorders resulting from defective diet or lack of exercise, the condition is frequently remedied by overcoming the cause. Lack of exercise and overfeeding seem to be etiological factors in a fair percentage of cases.

Besides abnormalities of the male genital tract, we must always consider the numerous factors in the female that may kill or weaken the sperms. Impediments to successful coitus may be present in the form of vaginal constrictions, abnormally short or small vagina, or other deformities. Hostile exudates, mechanical obstructions, and other factors may interrupt the progress of the sperms at any point in the tract.

Although little is known definitely regarding disorders of the endocrines in the bull and their relation to reproduction, the work in human medicine and experimental researches upon laboratory animals warrant thoughtful consideration of these factors which are by no means insignificant. In the future, these glands will no doubt receive more and more attention in their relation to the genital organs and reproduction.

## Conclusions

1. The genital organs of the bull quite frequently undergo pathological changes, due to infection with the same varieties of microorganisms associated with genital infections in the female.

2. In all probability, these microorganisms are frequently eliminated with the semen and infect the female during copulation.

3. Past or present infection in the genital organs of all the bulls so far examined was evidenced by the presence of the fine connective tissue tufts and strands upon the tunica vaginalis, particularly that part covering the tail of the epididymis.

4. Lowered sexual capacity is, as a rule, accompanied by demonstrable changes in the semen.

5. A study of impotency and sterility includes not only a thorough study of the genital organs, but also those extrinsic factors which govern reproduction either directly or indirectly.

6. A thorough knowledge of the anatomy and physiology of the male genital organs is fundamental to a clear understanding of the problem.

I am much indebted to Drs. W. L. and W. W. Williams for some of the material, and for helpful cooperation in the early part of the work; to Drs. C. M. Carpenter and R. R. Birch for many helpful suggestions; and to Dr. J. N. Frost and others who so kindly co-operated by placing at my disposal samples of semen and some of the genital tracts.

## Group III

| No. | Right seminal vesicle | Left seminal vesicle | Prostate | Cowper's | Right testis | Right epidid. | Left testis | Left epidid. | Right scrotal sac | Left scrotal sac |
|---|---|---|---|---|---|---|---|---|---|---|
| 1 | | | | | | | | | | |
| 2 | | Staph. alb. | | | | Staph. alb. Strep. vir. | | | | Strep. vir. |
| 3 | | | | | | | | | Strep. vir. | |
| 4 | Staph. alb. Strep. vir. | Staph. alb. Strep. vir. | | | | Staph. alb. Strep. haem. | | | Strep. vir. | Strep. vir. |
| 5 | | | | | | | | | | |
| 6 | | | | | | | | | | |
| 7 | | | | | Bact. short. | | | | | |
| 8 | Staph. alb. Strep. haem. | Staph. alb. | | | | | | Staph. alb. Strep. haem. | Strep. haem. | |
| 9 | | | | | | | | | | |
| 10 | Staph. alb. | | | | | | | | | |
| 11 | | | | | | | | | | |
| 12 | Staph. alb. | | | | Staph. alb. | | | Staph. alb. | | Strep. haem. |
| 13 | | | Staph. alb. | | | | | | | |
| 14 | | | | Staph. alb. | | | | | | |
| 15 | Staph. alb. | Staph. alb. | | | | Staph. alb. | Staph. alb. | | Strep. vir. | Strep. vir. |
| 16 | | | | | | | | | | |

Bacteriology of the Genital Tracts of Aborted Foetuses

## Group IV

| No. | Right seminal vesicle | Left seminal vesicle | Right testis | Left testis | Remarks |
|---|---|---|---|---|---|
| 1 | | | | | |
| 2 | Strep. vir. | | | Strep. vir. | Strep. vir. from heart's blood. |
| 3 | B. coli | B. coli | B. coli | B. coli | B. coli from all organs |
| 4 | | | Bact. short. | Bact. short. | Bact. abortum from spleen. |
| 5 | | | | | |
| 6 | | | | | |
| 7 | Bact. short. | | | Bact. short. | Bact. abortum from liver and heart's blood. |
| 8 | Staph. alb. | Staph. alb. | | Staph. alb. | Staph. alb. from heart's blood. |
| 9 | | | | | |
| 10 | Bact. short. | | Bact. short. | | Bact. abortum from heart's blood. |
| 11 | | | | | |

Bacteriology of the Genital Tracts of Calves Dying of Calf Dysentery

## Group V

| No. | Right seminal vesicle | Left seminal vesicle | Right testis | Left testis | Remarks |
|---|---|---|---|---|---|
| 1 | | | | | |
| 2 | B. coli Staph. alb. | B. coli Staph. alb. | | Staph. alb. | B. coli |
| 3 | | | | | |
| 4 | Strep. viridans | Strep. viridans | Strep. viridans | | Strep. viridans from heart's blood. |
| 5 | | | | | |
| 6 | | | | | |
| 7 | | | | | |

Bacteriology of the Genital Tracts of Matures Infected or Suprarenal Bulls

## Group VI

| No. | R.S.V. | L.S.V. | Prostate | Cowper's | R. Testis | R. Epidid. | L. Testis | L. Epidid. | R. scrotal sac | L. scrotal sac |
|---|---|---|---|---|---|---|---|---|---|---|
| 1 | Strep. vir. Staph. alb. | Staph. alb. | | | Strep. haem. Staph. alb. | Strep. vir. Staph. alb. | Staph. alb. | Strep. could not grow | | |
| 2 | Strep. (neutral) | Strep. vir. Staph. alb. | | | Staph. alb. | Strep. vir. Staph. alb. | Staph. alb. | Strep. vir. Staph. alb. | | |
| 3 | Strep. vir. | Strep. vir. Staph. alb. | | | | Strep. vir. Staph. alb. | Strep. vir. | | | Strep. vir. |
| 4 | | | | | Staph. alb. | Strep. vir. | | | | Strep. haem. |
| 5 | B. coli Staph. alb. | B. coli Staph. alb. | | | | Staph. alb. | | B. coli Staph. alb. | | |
| 6 | | Strep. vir. Staph. alb. | | | | | | Strep. vir. | | |
| 7 | | | | | | | | | | |
| 8 | Ps. pyocan. Strep. haem. | Ps. pyocan. Strep. haem. | | | | | | | | |
| 9 | Staph. alb. | Staph. alb. | Staph. alb. | | | Strep. vir. | Unident. rod | Unident. rod | | Strep. haem. |
| 10 | | | | | Staph. alb. | | | Staph. alb. | | |
| 11 | | Staph. alb. | | | Staph. alb. | Strep. haem. | | Strep. vir. Staph. alb. | Strep. (neutral) | |
| 12 | Staph. alb. | Strep. vir. | | Staph. alb. | | Strep. vir. | | | | |
| 13 | Staph. alb. B. coli | Staph. alb. B. coli | Staph. alb. | | | Strep. vir. | | | | |
| 14 | Staph. alb. | Staph. alb. | | | | Strep. (neutral) | | Strep. vir. | Strep. vir. | |
| 15 | | Strep. haem. | | | | | | Strep. haem. | | |
| 16 | Staph. alb. | Strep. vir. | | | | | Staph. alb. | Strep. vir. | | Strep. haem. |

A—Penicillin of the Genital Tracts of Matures Bulls in Internal Bulls

A—Pathology of the Genital Tract of Mature [illegible] Bulls

| | R. Sem. Vesicle | L. Sem. Vesicle | Ductus Deferens | R. Testis | R. Epididymis | L. Testis | L. Epididymis |
|---|---|---|---|---|---|---|---|
| 1 | Desquamation and degeneration of lining membrane. Debris and exudates in lumen. | Same as right. | Both apparently normal. | Apparently normal. | Acute desquamation and degeneration in some tubules. | Apparently normal. | Degeneration and desquamation in some tubules. Others atrophied due to a chronic interstitial inflammation. |
| 2 | Most vesicles degenerating. Interstitial cellular infiltration. Necrosis in one large area. | Same as right. | Right apparently normal. Left shows degeneration of the lining membrane. | Most tubules apparently normal. Some show degeneration of the seminal epithelium. | Degeneration and desquamation of epithelium of the tubules. | Most tubules becoming atrophied due to a chronic inflammation. | Same as right, but some tubules are becoming atrophied due to a chronic inflammation. |
| 3 | Entire gland degenerated. Coagulation necrosis of interstitial tissue. Debris and exudates in the vesicles. | Same as right. | Slight desquamation and degeneration of mucosa of right ductus. | Some tubules normal. Others show degeneration of the seminal epithelium. | Much of the epithelium denuded. Debris and exudates in the lumen. | Same as right. | Same as right. |
| 4 | Same as No. 3. | Same as No. 3. | Epithelium entirely denuded in both ducts. | Much degeneration and desquamation of the seminal epithelium. | Desquamation of most of lining membrane. | Same as right. | Same as right. |

B—Pathology of the Genital Tract of Mature [illegible] Bulls

| | R. Sem. Vesicle | L. Sem. Vesicle | Ductus Deferens | R. Testis | R. Epididymis | L. Testis | L. Epididymis |
|---|---|---|---|---|---|---|---|
| 5 | Epithelium becoming degenerated. Vesicles filled with debris and exudates. | Same as right. | Both apparently normal. | Degeneration of many tubules. | All epithelium denuded. Debris in lumen. | Same as right. | Nearly normal. |
| 6 | Acutely inflamed. | Same as right. | Slight degeneration of lining membrane of both. | Degeneration of mucosa in some tubules. | Apparently normal. | Degeneration in some tubules. No mitosis in some. | Apparently normal. |
| 7 | Some degeneration of lining membrane. | Same as right. | Both apparently normal. | Apparently normal. | Apparently normal. | Apparently normal. | Apparently normal. |
| 8 | Necrosis in areas. Infiltration of interstitial tissue. Degeneration of lining membrane. | Same as right. | Both apparently normal. | Apparently normal. | Apparently normal. | Apparently normal. | Apparently normal. |

C—Pathology of the Genital Tract of Mature [illegible] Bulls

| | R. Sem. Vesicle | L. Sem. Vesicle | Ductus Deferens | R. Testis | R. Epididymis | L. Testis | L. Epididymis |
|---|---|---|---|---|---|---|---|
| 9 | Apparently normal. | Apparently normal. | Both apparently normal. | Apparently normal. | Apparently normal. | Apparently normal. | Apparently normal. |
| 10 | Slight desquamation and degeneration of the lining membrane. | Same as right. | Both apparently normal. | Apparently normal. | Degeneration of lining membrane. | Apparently normal. | Same as right. |
| 11 | Apparently normal. | Apparently normal. | Both apparently normal. | Some tubules show no mitosis, others are degenerating. | Degeneration and desquamation of lining membrane. | Same as right. | Degeneration and desquamation of lining membrane. |
| 12 | Degeneration and desquamation of epithelium of some vesicles. | Same as right. | Both apparently normal. | Apparently normal. | Degeneration of epithelium of all tubules. Debris and exudates in lumen. Some increase in connective tissue. | Apparently normal. | Apparently normal. |

D—Pathology of the Genital Tract of Mature [illegible] Bulls

| | R. Sem. Vesicle | L. Sem. Vesicle | Ductus Deferens | R. Testis | R. Epididymis | L. Testis | L. Epididymis |
|---|---|---|---|---|---|---|---|
| 13 | Degeneration of most of gland. Cellular infiltration of interstitial connective tissue. Debris and exudates in lumen. | Same as right. | Slight catarrhal inflammation of mucosa of both. | Apparently normal. | Degeneration and desquamation of lining membrane of most tubules. | Apparently normal. | Same as right. |
| 14 | Apparently normal. | Apparently normal. | Apparently normal. | Some degeneration of seminal epithelium of some tubules. | Apparently normal. | Apparently normal. | Acute inflammation. Some desquamation of lining membrane. |
| 15 | Slight desquamation and degeneration of lining membrane. | Acute catarrhal inflammation. Debris and exudates in lumen. | Both apparently normal. | Degeneration of seminal epithelium in some tubules. | Desquamation of lining membrane. Debris in lumen. | Apparently normal. | Apparently normal. |
| 16 | Chronic interstitial inflammation. Atrophy of many tubules. | More acute than right. | Both apparently normal. | Degeneration in some tubules. Some show poor mitosis. | Apparently normal. | Same as right. | Acute inflammation. Some desquamation and degeneration of lining membrane. |

# BIBLIOGRAPHY

1.

BANG B. Die Aetiologie des seuchenhaften (infectiösen) Verwerfens. Zeit Med. Band I, S. 241, 1897. Das seuchenhaften Verwerfen der Rinder. Arch. wiss. u. prakt. Thierheilk., Band 33, S. 312, 1907.

2.

LAW, JAMES. Contagious Abortion of Cows. Circular No. 5, Agr. Experiment Sta. University of California, June, 1903.

3.

Report of English Commission of Epizootic Abortion. Appendix to Part I, p. 17, 1909.

4.

HADLEY, F. B., and LOTHE, H. The Bull as a Disseminator of Contagious Abortion. Jour. Amer. Vet. Med. Assoc., L, 1916–17, p. 143.

5.

HADLEY, F. B. Contagious Abortion Questions Answered. Bulletin No. 296, Oct. 1921, Agr. Experiment Sta., University of Wisconsin.

6.

BUCK, J. M., CREECH, G. T., and LADSON, H. H. Bacterium Abortus Infection of Bulls (Preliminary Report). Jour. Agr. Research, August, 1919.

7.

SCHROEDER, E. C., and COTTON, W. E. The Bull as a Factor in Abortion Disease.

8.

COTTON, W. E. Proceedings of A. V. M. A., p. 851, 1913.

9.

CARPENTER, C. M. Report of the New York State Veterinary College, Cornell University, 1920–21.

10.

RETTGER, L. F., and WHITE, G. C. Infectious Abortion in Cattle. Storrs Agr. Exp. Sta. Bulletin No. 93. January, 1916.

11.

McFADYEAN, SHEATHER and MINETT. Researches Regarding Epizootic Abortion. Jour. of Comp. Path. and Therap., XXVI, 142, 1913.

12.

SCHROEDER, E. C. Bureau of Animal Industry Investigations of Bovine Infections Abortion. Jour. Amer. Vet. Med. Assoc., LX, p. 542. February, 1922.

13.

BARNEY, J. D. Recent Studies on the Pathology of the Seminal Vesicles. Bost. Med. and Surg. Jour., CLXXI, 1914, 59.

14.

WILLIAMS, W. L. The Diseases of Bulls. Cornell Veterinarian. X, 94. April, 1920.

15.

WILLIAMS, W. L. Report of the New York State Veterinary College, Cornell University, 1920–21.

16.

WILLIAMS. W. W. Technique of Collecting Semen for Laboratory Examination with Review of Several Diseased Bulls. Cornell Vet. X, 87, April, 1920.

17.

WILLIAMS, W. W. Diseases of the Bull Interfering With Reproduction. Jour. of Amer. Vet. Med. Assoc., LVIII, 29, October, 1920.

18.

HOPPER, E. B. Herd Efficiency from the Standpoint of the Veterinarian. North Amer. Vet., III, 71, February, 1922.

19.

WILLIAMS, W. W. Observations on Reproduction in a Purebred Dairy Herd. Cornell Veterinarian, XII, 19, January, 1922.

20.

WALKER, K. M. The Diagnosis and Treatment of Sterility in the Male. Lancet, CCI, 228. July 30, 1921.

21.

PENDE. Endocrinoglia-Pathologia E Clinica. Abstract in Endocrinology, II, 42.

22.

FERGUSON, J. S. Normal Histology and Microscopical Anatomy.

23.

ELLENBERGER. Vergl. mikroskop. Anatomie der Haustiere. II.

24.

COURRIER, M. R. On the Existence of a Secretion of the Epididymis of the Hibernating Bat and Its Significance. C. R. Soc. de Biol. (Paris), 1920, LXXXIII, 67–69.

25.

STIGLER, R. Abstract in Jour. Phys. et Path. Gen. (2) CLXXI, 273, September, 1918.

26.

DISSELHORST. Ansfuhrapparat und Anhangsdrusen der männlichen Geschlechtsorgane. Oppel's Lehrbuch der vergleichenden mikr. Anat. der Wirbeltiere, IV, Jena, 1904.

27.

RYTINA, A. A. The Verumontanum, With Special Reference to the Sinus Pocularis: Its Histology, Anatomy, and Physiology. Jour. Urology, I, 1917, 231.

28.

FISH, P. A. The Spermatic Secretion and Its Ultraparticles. Cornell Veterinarian, October, 1921.

29.

MARSHALL, F. H. A. The Physiology of Reproduction.

30.

KINGSBURY, B. F. Professor of Histology and Embryology, Cornell University, Ithaca, New York.

31.

WALKER. Arch. f. Anat. u. Entwicklungsgesch., 1899, und Arch. f. Anat. u. Physiol. 1899.

32.

BOETTCHER, W. On the Significance of the Secretions of the Male Accessory Genital Organs. Münch. med. Wchnschr. 1920, 67, 1, p. 44.

33.

EDITORIAL. Jour. of Amer. Med. Assoc., LXXVII, No. 1, 42, July 2, 1921.

34.

REYNOLDS, E. Fertility and Sterility. Jour. of Amer. Med. Assoc., LXVII, 1193–1199, October 21, 1916.

35.

CARY, W. H. Examination of Semen With Special Reference to Its Gynecological Aspects. Amer. Jour. of Obstetrics and Diseases of Women and Children. LXXIV, No. 4, 1916.

36.

WOLF, C. G. L. The Survival of Motility in Mammalian Spermatozoa. Jour. of Physiol. IV, 246, August 3, 1921.

37.

BROWN, J. H. The Use of Blood Agar for the Study of Streptococci. Monograph of the Rockefeller Institute for Med. Res., 1919, IX.

38.

GARNETT, J. B., and others. The Surgical Treatment of Sterility due to Obstruction at the Epididymis Together with a Study of the Morphology of Human Spermatozoa. Univ. of Penn. Med. Bulletin, March, 1902.

39.

REYNOLDS, E. and MACOMBER, D. Defective Diet as a Cause of Sterility. Jour. of Amer. Med. Assoc., LXXVII, 169, July 16, 1921.

40.

NOVARRO, P. A. Tissues of the Testicle and Antaminosis. Gazetta degli Ospedali (Milano), 1920, XLI, 424.

41.

WILLIAMS, W. L. Observations on Reproduction in a Pure Bred Beef Herd. Cornell Veterinarian. January, 1922.

42.

BELL, W. B. Correlation of Function: With Special Reference to the Organs of Internal Secretion and the Reproductive System. Brit. Med. Jour., 1920, 1, 787.

43.

BROWN, W. L. The Principles of Internal Secretion. Brit. Med. Jour, 1920, II, 687–691.

44.

JUMP. Discussion in Penn. Med. Jour., XXV, 81.

45.

BIEDL, A. The Internal Secretory Organs.

46.

HEWER, E. E. The Effect of Thymus Feeding on the Activity of the Reproductive Organs of the Rat. Jour. of Physiology, XLVII, 1913–1914, 479.

47.

LLOYD-JONES, O. and HAYS, F. A. The Influence of Excessive Sexual Activity of Male Rabbits. 1. On the Properties of the Seminal Discharge. Jour. of Exper. Zoology, 1918, XXV, 463.

48.

WILLIAMS, W. L. Improvement of the Reproductive Efficiency of Cattle. North Amer. Veterinarian, III, May, 1922.

## DESCRIPTION OF PLATES

PLATE I.

Fig. 1.

Diagrammatic sketches showing the development of a spermatozoon from a spermatogonium.

Fig. 2.

Diagrammatic sketch showing the minute structure of a spermatozoon. The middle piece is made comparatively thick in order to bring out the finer structures. Adapted from Ellenberger.

PLATE II.

Fig. 3.

Testicle of bull, showing extensive degeneration and necrosis.

PLATE III.

Fig. 4.

Inner part of wall of ductus deferens. Normal. × 230.

Fig. 5.

Same, but showing extensive degeneration and exfoliation of the lining membrane. × 230.

PLATE IV.

Fig. 6.

Ductus deferens, showing entire exfoliation of the lining membrane. The lumen is filled with a cellular debris. × 50.

Fig. 7.

Same, showing the degeneration of the lining membrane, and debris in lumen. × 230.

PLATE V.

Fig. 8.

Seminal vesicle of bull. High power section showing the normal structure of the vesicular cavities.

Fig. 9.

Seminal vesicle of bull. Low power. The membrane lining the cavities is degenerated and exfoliated. The cavities are filled with cellular debris, and exudates. There is some increase in the interstitial tissue, and atrophy of some of vesicular cavities.

PLATE VI.

Fig. 10.

Low power section of seminal vesicle of bull. The condition is about the same as in Fig. 9 except that it is not quite as severe.

Fig. 11.

Same as Fig. 10. High power.

PLATE VII.

Fig. 12.

Testicular tubule showing normal spermatogenesis. × 230.

Fig. 13.

Testicular tubule showing no evidence of mitosis. × 230.

Fig. 14.

Testicular tubule. The spermatogenic epithelium is beginning to degenerate and become cast off into the lumen. × 230.

Fig. 15.

Testicular tubule, showing almost total exfoliation of the spermatogenic epithelium. × 230.

PLATE VIII.

Fig. 16.

Testicular tubule. The seminal epithelium is entirely degenerated. The membrana propria is markedly thickened. × 230.

Fig. 17.

Testicular tubule. The tubule is undergoing atrophy and degeneration. The interstitial connective tissue is much increased in amount. × 230.

Fig. 18.

Same as Fig. 17, except that it is in the more advanced stages.

Fig. 19.

Testicular tubule. The spermatogenic epithelium has undergone a sort of hydropic degeneration. The interstitial connective tissue has become much increased in amount and has undergone cellular infiltration.

PLATE IX.

Fig. 20.

Normal structure of epididymis tubule. × 140.

Fig. 21.

Epididymis tubule showing exfoliation of the lining membrane, and cellular debris in the lumen. × 140.

Fig. 22.

Atrophy and degeneration of epididymis tubule. The interstitial connective tissue is much increased in amount. × 140.

Fig. 23.

Same, but in more advanced stage. × 140.

PLATE X.

Fig. 24.

About the same as Fig. 22. × 140.

Fig. 25.

Marked degeneration of epididymis tubule. There is a cellular infiltration of the interstitial tissue. × 60.

PLATE XI.

Fig. 26.

Spermatozoon, showing constriction at middle of head. The head is also somewhat contorted. × 670.

Fig. 27.

Spermatozoon. Pear shaped head. × 670.

Fig. 28.

Spermatozoon. The head is quite long, and pointed at its posterior end. × 670.

Fig. 29.

Spermatozoon, showing a constriction at middle of the head. × 670.

PLATE XII.

Fig. 30.

Microcephalic spermatozoon. × 670.

Fig. 31.

Spermatozoon. The head is small, and pear shaped. × 670.

Fig. 32.

Macrocephalic sperm. The middle piece is much thickened. × 670.

Fig. 33.

Tailless spermatozoa. × 670.

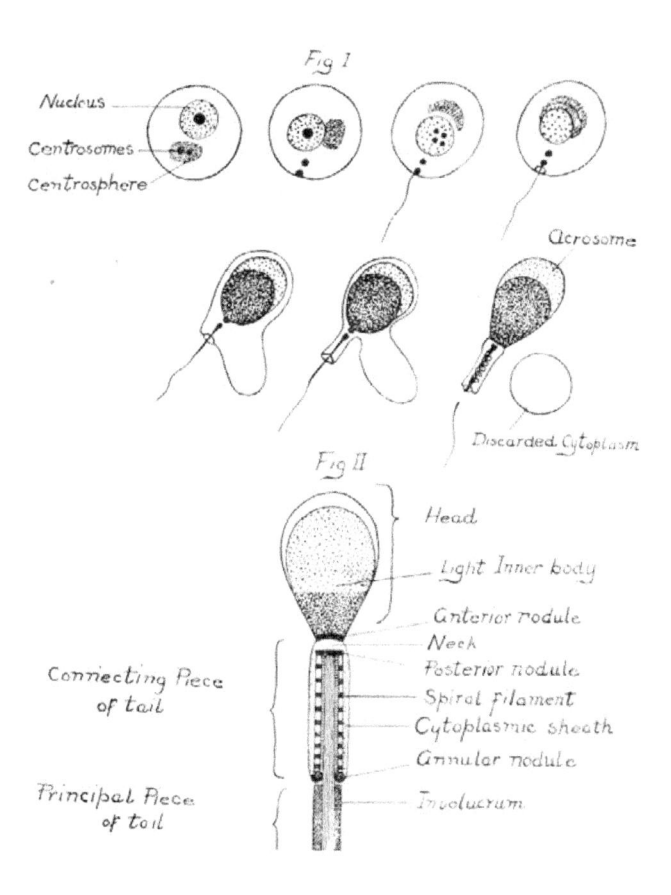

PLATE I

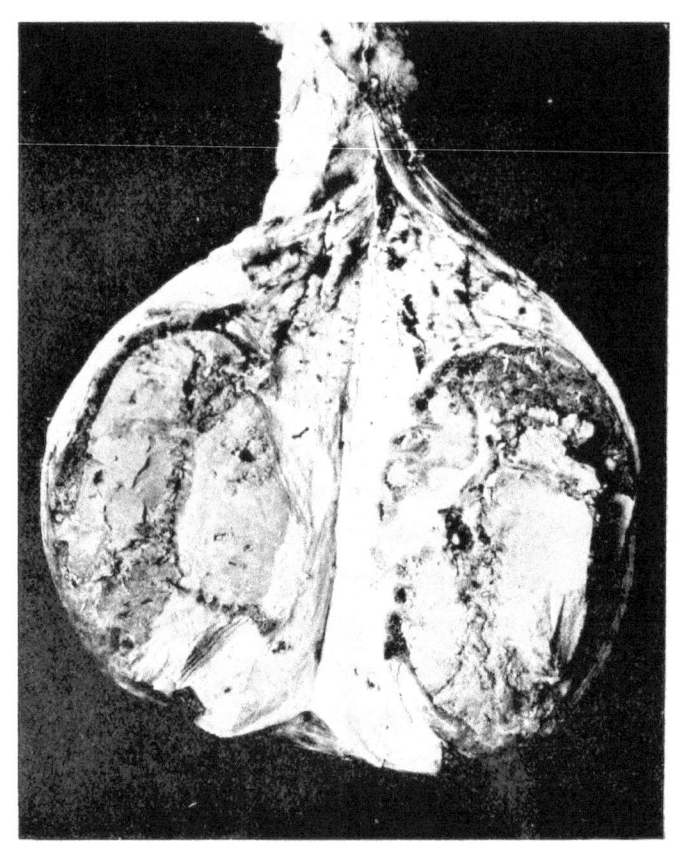

Fig. 3.

PLATE II

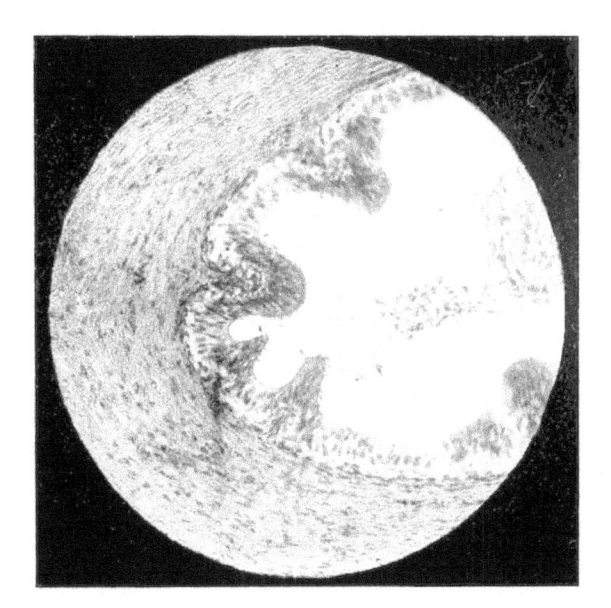

Fig. 4.

Fig. 5.

PLATE III

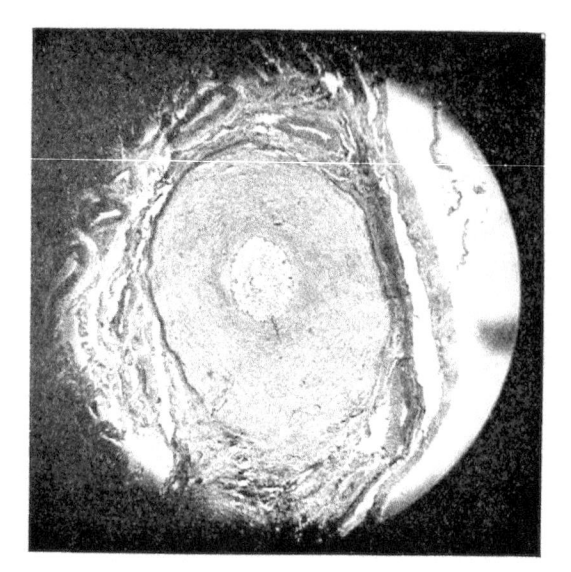

Fig. 6.

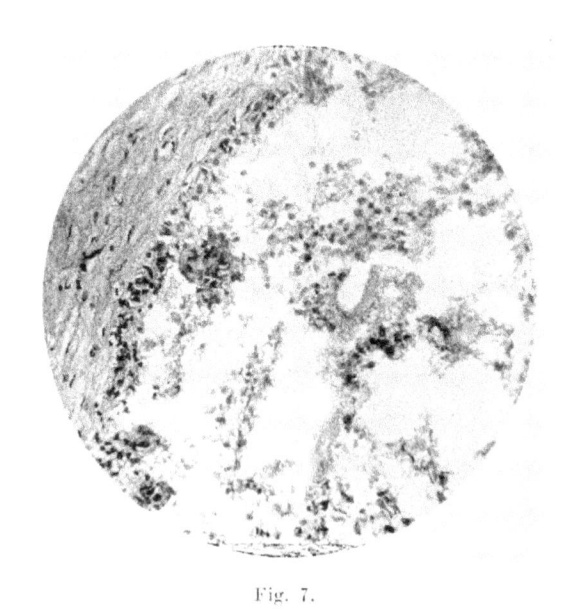

Fig. 7.

PLATE IV

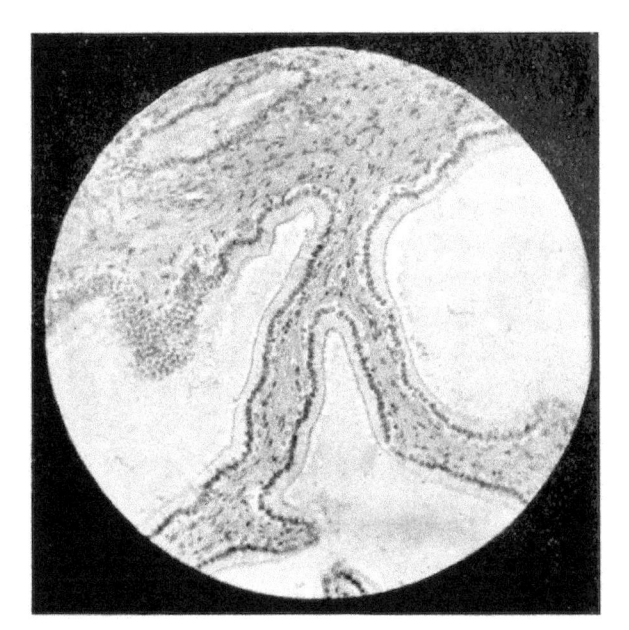

Fig. 8.

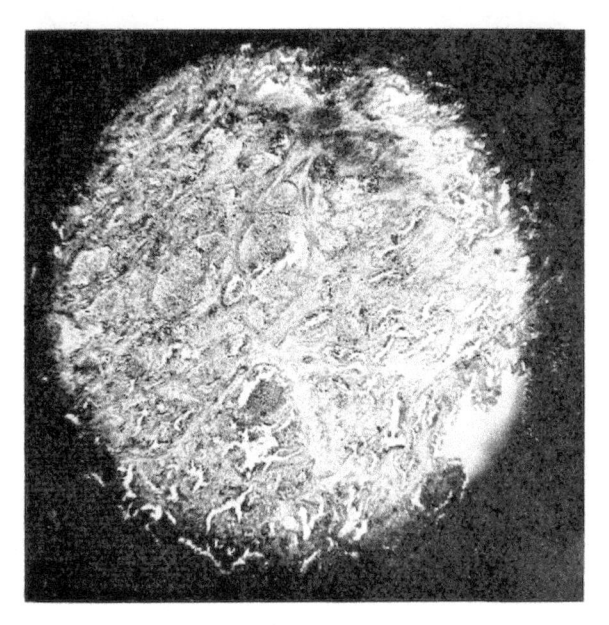

Fig. 9.

PLATE V

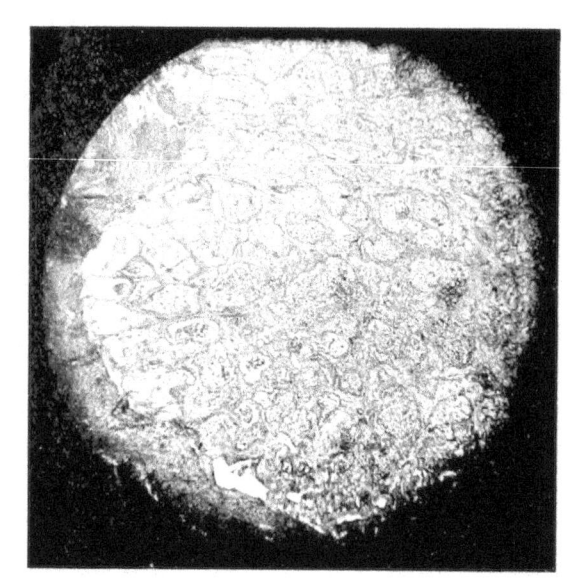

Fig. 10.

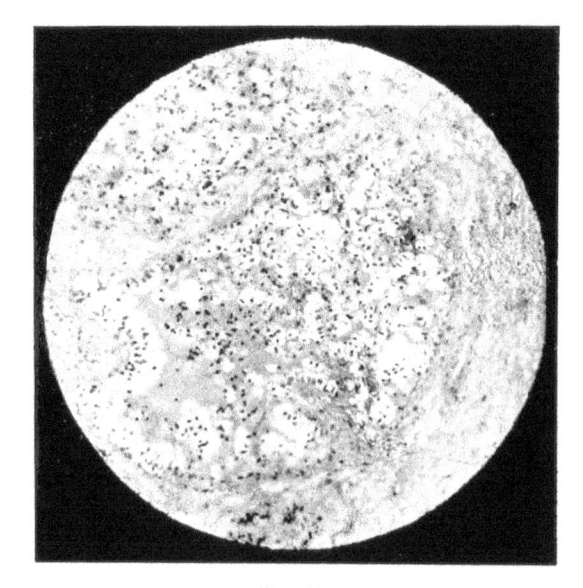

Fig. 11.

PLATE VI

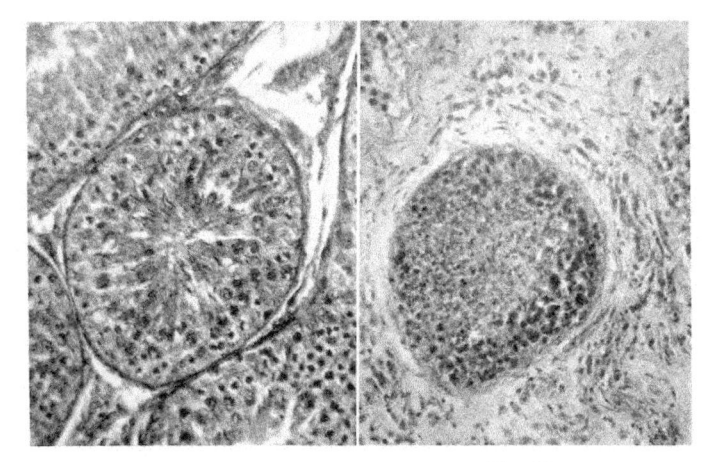

Fig. 12.    Fig. 13.

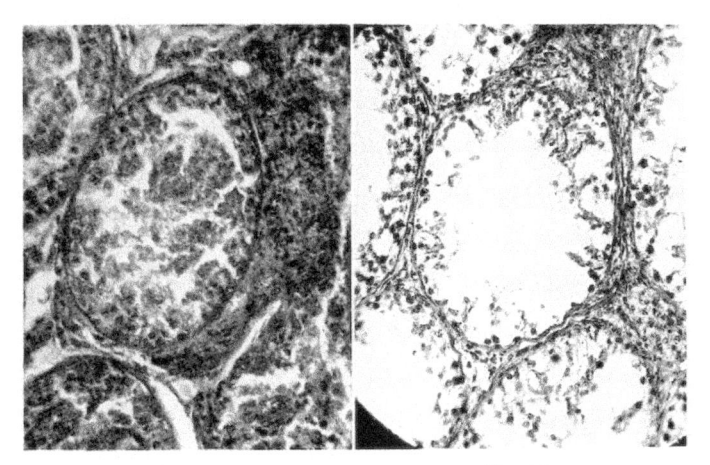

Fig. 14.    Fig. 15.

PLATE VII

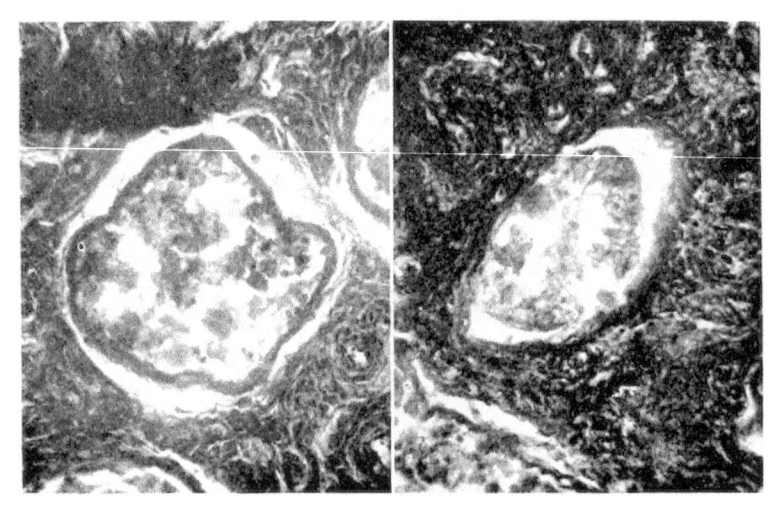

Fig. 16.  Fig. 17.

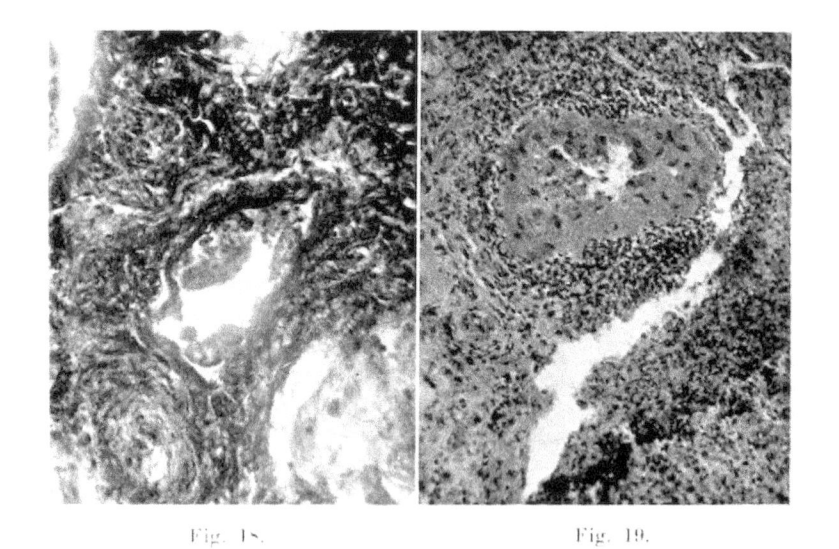

Fig. 18.  Fig. 19.

PLATE VIII

- 70 -

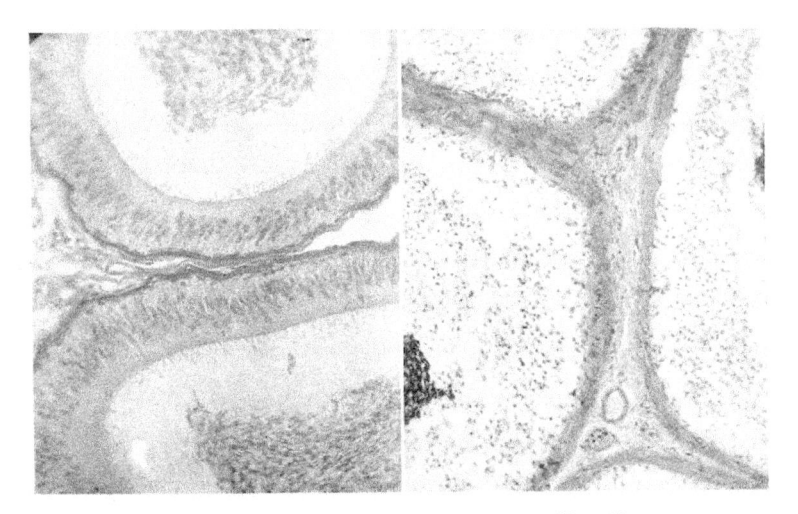

Fig. 20.

Fig. 21.

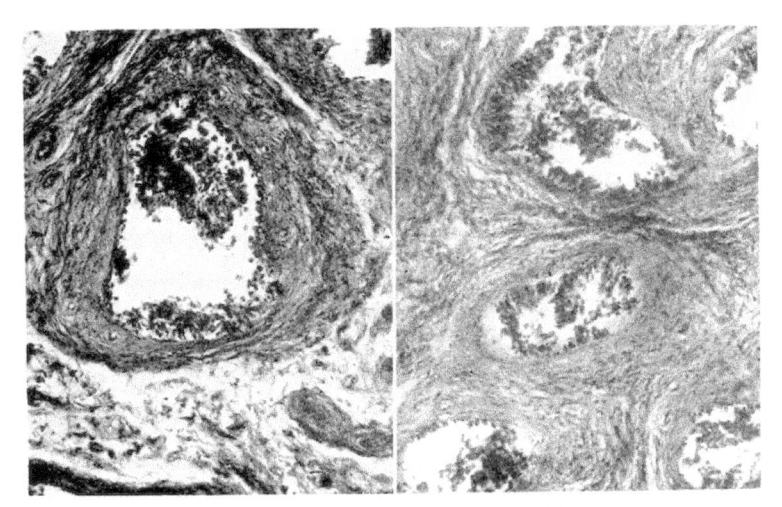

Fig. 22.

Fig. 23.

PLATE IX

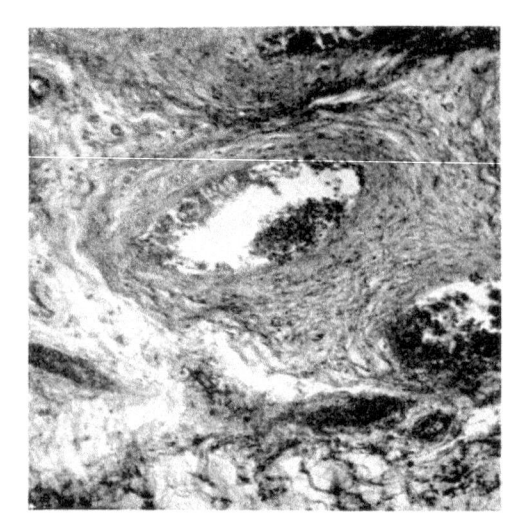

Fig. 24.

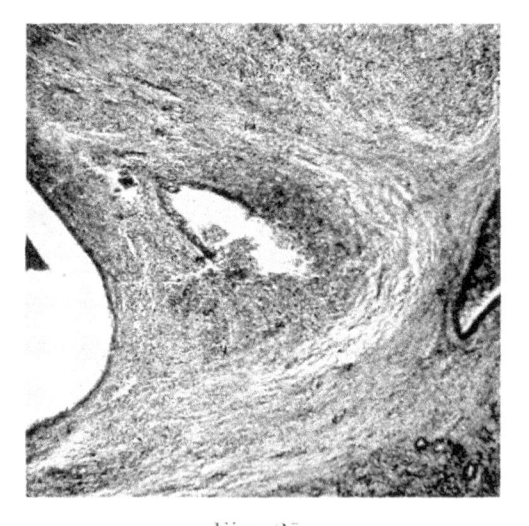

Fig. 25.

PLATE X

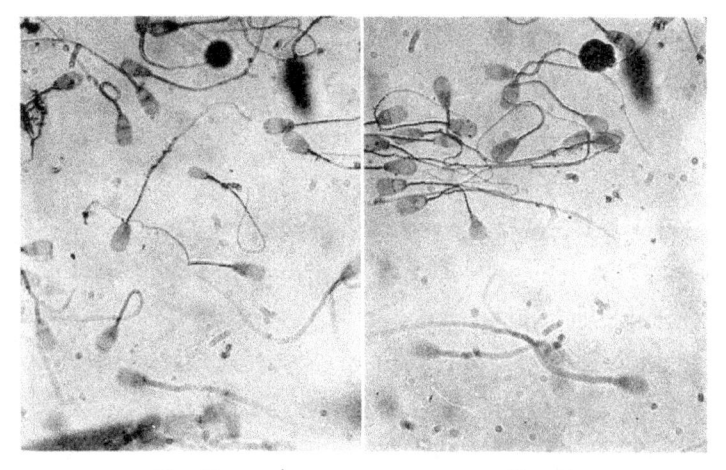

Fig. 26.    Fig. 27.

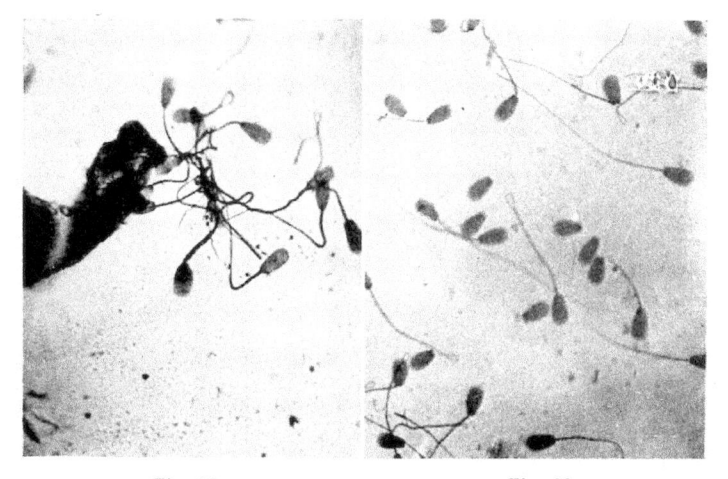

Fig. 28.    Fig. 29.

PLATE XI

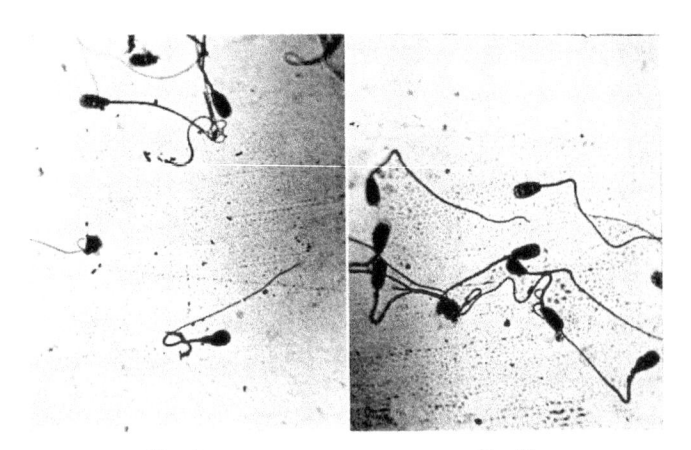

Fig. 30.  Fig. 31.

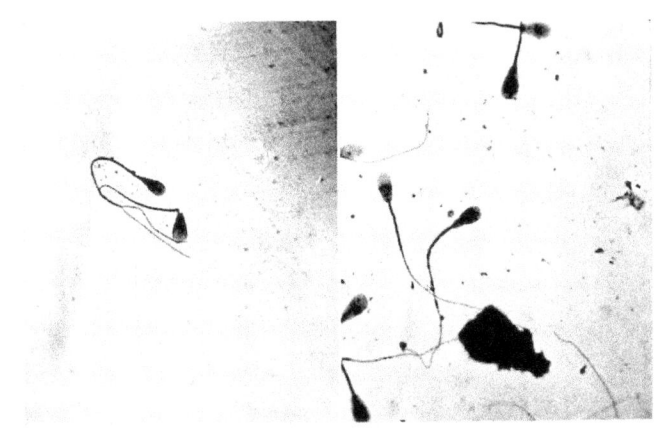

Fig. 32.  Fig. 33.

PLATE XII